Intermediate
GEOGRAPHY

Calvin Clarke

Hodder & Stoughton

A MEMBER OF THE HODDER HEADLINE GROUP

The publishers would like to thank the following individuals, institutions and companies for permission to reproduce photographs in this book. Every effort has been made to trace ownership of copyright. The publishers would be happy to make arrangements with any copyright holder whom it has not been possible to contact.

Abbott Mead Vickers BBDO Ltd (62.2); © Shaen Adey/Gallo Images/CORBIS (57.2); Argus Fotoarhiv GmbH (50.2); © Yan Arthus-Bertrand/CORBIS (82.1); Associated Press (82.4, 85.1); William Bagshaw, White Scar Caves (14.4); Gary Braasch/CORBIS (73.4); © Carol Cohen/CORBIS (73.5); © CORBIS (74.3, 74.4); Durlston Country Park (23.2); Mark Edwards/Still Pictures (88.3); Eye Ubiquitous (54.1); Eye Ubiquitous/CORBIS (74.2); Owen Frankin/CORBIS (78.3); GeoScience Features Picture Library (5.2, 5.3, 6.1, 11.3, 13.4, 17.1, 17.2, 19.3, 24.2, 24,3, 25.2, 26,2); Tim Grevatt LBIPP, Figurehead Photographic (13.5); © Annie Griffiths Belt/CORBIS (70.1); © Bob Krist/CORBIS (45.2); Emma Lee (41.3, 45.1); © Michael Nicholson/CORBIS (48.1); John Noble (4.2, 13.2, 13.3); © Tim Page/CORBIS (32.3); © Caroline Penn/CORBIS (64.2, 66.2, 66.3); Richard Powers (53.1); Chris Rainier (66.4); © Roger Ressmeyer/CORBIS (71.3, 76.4); © Reuters New Media Inc/CORBIS (77.4); Nigel Shuttleworth (27.2); David Simpson (32.2, 37.2, 48.3, 49.2, 52.1, 52.3); Skyscan Photolibrary/Bob Evans (20.3); © Liba Taylor/CORBIS (88.1); Andrew Ward (23.3, 57.1); © Patrick Ward/CORBIS (14.3); Wellcome Trust Photo Library (61.2, 63.1, 65.1); © Nick Wheeler/CORBIS (78.1, 78.2); © Jim Winkley, Ecoscan (15.1); © Adam Woolfit/CORBIS (14.5)

The illustrations were drawn by Peters & Zabransky Limited.

Figures 7.1, 15.2, 21.2: Maps reproduced from Ordnance Survey mapping with the permission of the Controller of Her Majesty's Stationery Office, © Crown Copyright

Orders: please contact Bookpoint Ltd, 130 Milton Park, Abingdon, Oxon OX14 4SB. Telephone: (44) 01235 827720. Fax: (44) 01235 400454. Lines are open from 9.00 – 6.00, Monday to Saturday, with a 24 hour message answering service. Email address: orders@bookpoint.co.uk

British Library Cataloguing in Publication Data
A catalogue record for this title is available from the British Library

ISBN 0 340 77533 5

Published by Hodder & Stoughton Educational Scotland
First Published 2001
Impression number   10 9 8 7 6 5 4 3 2
Year   2007 2006 2005 2004 2003 2002

Copyright © 2001 Calvin Clarke

Cover photo from Photodisc
Typeset by Fakenham Photosetting Ltd
Printed in Italy for Hodder & Stoughton Educational, a division of Hodder Headline Plc, 338 Euston Road, London NW1 3BH.

Introduction

This book has been written to cover the Scottish National Courses: Geography Intermediate 1 and 2. These courses are very similar. The Intermediate 2 course includes all the content of the Intermediate 1 syllabus, together with some additional content. Both courses are divided into the same three Units, and each Unit is split into two Topics. Students must study one Topic from each Unit. The titles of the Topics and Units are given below.

Unit 1—Scotland/British Isles

Topic (a) **Physical Landscapes and Land Use**
Topic (b) **Landscapes and Tourism**

Unit 2—Europe

Topic (a) **Environmental Issues**
Topic (b) **Population**

Unit 3—Global Issues

Topic (a) **Development and Health**
Topic (b) **Environmental Hazards**

Both Topics from each Unit are covered in this book. This allows Intermediate 1 students in the Fifth Year to study the alternative Topic in the Sixth Year, should they wish to go on to study Intermediate 2.

Both Topics in Unit 1 include a study of glaciated uplands. Therefore, the first ten chapters of this book should be studied by everyone, irrespective of whether they have opted for Topic (a) or Topic (b).

There is some choice within both Intermediate syllabi and the Intermediate 1 syllabus is shorter than that of Intermediate 2. This means that students do not need to study every chapter within their selected Topics. The list below shows the choices open to students and their teachers.

Intermediate 1 Course

Unit 1, Topic (b): students do not need to study volcanic landscapes (chapters 24–29).

Unit 2, Topic (b): students need to learn about *either* Berlin (48–50) *or* Paris (53–54) and they need only learn about *either* Brandenburg (51–52) *or* the Paris Basin (55–56).

Unit 3, Topic (a): students should cover *either* heart disease (67–68) *or* cancer (69–70), and they should also cover *either* malaria (63–64) *or* cholera (65–66).

Unit 3, Topic (b): students must know a case study of *either* a volcanic eruption (73–74) *or* an earthquake (77–78), and they need to know a case study of *either* a tropical storm (81–82) *or* a flood (84–85) *or* a drought (87–88).

For Unit 2, Topic (a) the course only requires the student to be familiar with case studies of two European landscapes under environmental pressure. However, it is still necessary for them to study all the chapters (30–42) in order to know the different types of landscape under pressure and their locations, and to understand the sources of the pressures.

Intermediate 2 Course

Unit 3, Topic (b): students must learn case studies of volcanic eruptions and earthquakes (chapters 73, 74, 77, 78) but they need only learn case studies of *two of the three weather hazards*—a tropical storm (81–82), a flood (84–85) or a drought (87–88).

★★For Unit 2, Topic (a) the course only requires

the student to be familiar with case studies of four European landscapes under environmental pressure. However, it is still necessary for them to study all the chapters (30–42) in order to know the different types of landscape under pressure and their locations, and to understand the sources of the pressures.

The Intermediate course requires the students to be skilled in several geographical methods and techniques. These are developed throughout the book and are duplicated in each of the two Topics so that students practise the same skills irrespective of the Topic selected.

The Topics are divided into double-page chapters, and each double-page includes a set of questions. Most of the questions are similar to those asked in the external examination and are a mixture of Intermediate 1 and 2 questions. One question in each double-page is also pitched at

Access 3 level for the benefit of anyone experiencing difficulty in tackling the others. These questions are 'starred' on the black and white spreads and numberless on the colour spreads. Generally, questions worth 4–6 marks are at Intermediate 2 level whereas those worth 1–3 marks are Intermediate 1 questions. Although a precise measure of a student's attainment is not offered, it is thought than an overall score of 50% or higher for a set of questions would be approximately equivalent to Intermediate 2 level and it should also be possible to monitor each student's progress by comparing their percentage marks for each set of questions they answer.

To assist students having difficulty in reading and learning all the text on each double-page, the key phrases are given in bolder type. These may also help all students when they revise for their internal assessments and external examinations.

Contents

Unit 2: Europe

Topic (a): Environmental Issues

Topic (b): Population

Unit 3: Global issues

Topic (a): Development and Health

Topic (b): Environmental Hazards

1 The shaping of the British Isles (1)

The British Isles millions of years ago

In Western Australia it is possible to travel for hundreds of kilometres across an unchanging landscape of flat featureless plains. In northern Canada and Russia you can travel for the same distance through vast tracts of coniferous forests. But, within a few hundred kilometres in the British Isles, the physical landscape changes from steep, rugged mountains to flat plains, from woodland and moorland to bare rock plateaux, from wide river valleys to deep gorges and, on the coast, from long tracts of sandy beaches to dramatic cliff headlands and wide sea inlets.

To find out why we have such landscapes we must go back in time—a long, long way back in time. And if we go back far enough we find that the British Isles used to have landscapes as dramatic as any found in the world today.

Figure 1.1 One million years ago almost everywhere was covered in ice

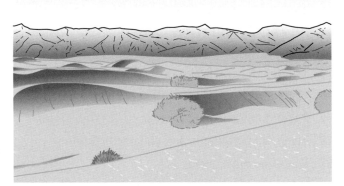

Figure 1.2 250 million years ago some of the British Isles was under the sea whereas the rest was a hot desert

Figure 1.3 330 million years ago our climate was hot and wet with tropical rainforests growing

Figure 1.4 450 million years ago we had mountains higher than any in the world today

Figures 1.1 and 1.4 provide clues as to what has shaped our land. The most awesome force has been mountain building. Crustal plates crunching together long ago squeezed up rocks into mountains and, as all the rocks were stretched, cracks appeared and millions of tonnes of lava poured out from the cracks. This now forms distinctive areas of high land, such as Mt. Snowdon in Wales, Edinburgh's Castle Rock and Arthur's Seat and the Cuillin Hills on Skye.

Agents of erosion

From the time these mountains were formed they have been weathered and eroded so much that only their stumps remain today. In a few million years they will have been worn away completely.

The wearing down of rocks has been carried out chiefly by the big four agents of erosion—wind, waves, rivers and moving ice. They have not only eroded vast amounts of rock, they have also transported it away and then deposited it, creating new landforms as they did so.

Figure 1.5 Erosion by wind

The power of the wind picks up sand and smaller particles and blasts them against rocks, slowly wearing them away. **They deposit the material they have eroded in the form of dunes,** when they slow down and lose power.

Moving ice freezes onto rocks and then tears them away as it moves forward. It deposits this material in the form of hummocky mounds of soil when the ice finally melts.

Figure 1.6 Erosion by moving ice

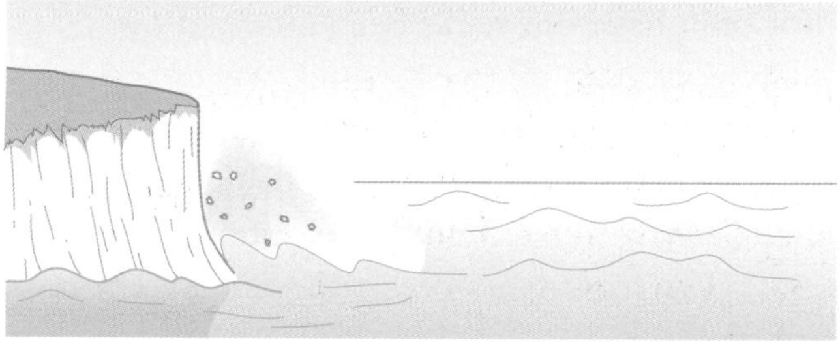

Figure 1.7 Erosion by moving water

Fast-flowing rivers and large waves have enough energy to pick up large rocks and boulders, which they use to pound against the rocks, breaking them up. Then, the sheer power of the moving water removes them. They are deposited in the form of mudbanks and beaches when they eventually slow down and lose energy.

QUESTIONS

1 Name the four main agents of erosion (2)

2 Describe the process by which wind erodes (2)

3 What happens to the material eroded by the wind? (1)

4 Describe the processes by which moving water erodes (4)

5 What happens to the material deposited by moving water? (1)

Put these agents of erosion into order, according to how powerful you think they are—wind, rivers, moving ice and waves. Give reasons for your answer (5)

2 The shaping of the British Isles (2)

Agents of weathering

Not only have our rocks been attacked for millions of years by rivers, wind, waves and moving ice, they have also been worn down by the weather. Over long periods of time rocks have been weathered until they crumble or rot away. The main types of weathering are chemical and physical.

Rain falls on limestone rock and seeps into its cracks

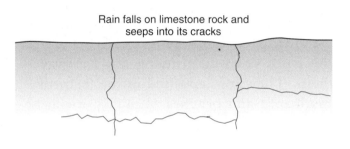

Acids in the rain dissolve the limestone, making the cracks wider

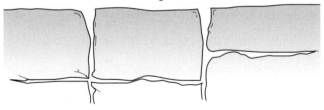

Gradually, the surface of the rock becomes lower

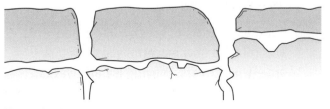

Figure 2.1 Chemical weathering

Chemical weathering — rain action

Rain is a weak acid. This is because it absorbs carbon dioxide as it falls through the atmosphere. When it reaches the Earth's surface, **it dissolves some of the minerals that make up rocks**, especially calcium carbonate found in limestone. The rain passes down into the rock through cracks, dissolving the rock and leaving potholes and caves, which eventually collapse (Figure 2.1).

Physical weathering — freeze–thaw action (frost action)

Rainwater enters cracks in rocks and, in winter, this water sometimes freezes. **When it freezes it expands, which forces the crack wider.** When this is repeated thousands of times **pieces of rock break off** (see Figure 2.2).

Physical weathering — onion-skin weathering

On very hot days **the outer surface of the rock heats up and expands. When it cools down at night it contracts** or shrinks. When this is repeated many, many times the surface cracks and **pieces of rock begin to break off.**

Rock types

Our landscape is shaped by the forces of weathering and erosion but the amount of weathering and erosion that takes place depends very much upon the type of rock underneath. In some areas the rock type is resistant (e.g. granite) and is eroded very slowly whereas, elsewhere,

Rain fills a crack in a rock

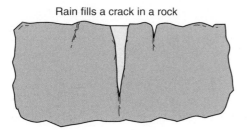

The water freezes and expands
and the crack is made wider

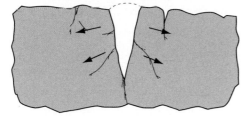

Eventually, the rock breaks up

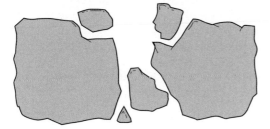

Figure 2.2 Physical weathering: freeze-thaw action

rocks such as clay and shale are much less resistant and are eroded quickly. In addition some rocks such as limestone are made of calcium and suffer badly from chemical weathering. Others have many cracks in them and are subject to a lot of freeze–thaw weathering.

Rocks are divided into three categories according to the way in which they formed.

Sedimentary rocks are made up of sediments laid down in layers, usually at the bottom of seas and oceans. The sediments have been eroded and weathered from other rocks. Examples include sandstone, limestone, chalk and clay. They are mostly softer rocks and usually form low land.

Igneous rocks have formed from molten rock, which has risen from the mantle through cracks in the Earth's crust and then cooled down

and solidified. Examples include granite, basalt and dolerite. These are all resistant rocks and almost always form high land.

Metamorphic rocks begin life as sedimentary or igneous rocks, but **are then changed by being subject to intense heat or pressure**. Examples include marble (from limestone), slate (from shale) and gneiss (from granite). Metamorphic rocks are also resistant rocks, which usually form high land.

All the agents of erosion and weathering are more powerful in uplands than lowlands. So they wear away our uplands but tend to deposit material in the lowlands. As a result our landscape is becoming flatter and, in time, may become completely flat.

QUESTIONS

1 Name two types of physical weathering and one type of chemical weathering　　(2)

2 Describe how rain can weather rocks　　(3)

3 Explain how sedimentary rocks have been formed　　(2)

4 What is the difference between an igneous and a metamorphic rock?　　(2)

5 Describe the differences in hardness between the three major rock types　　(2)

6 In what way does rock type affect the height of the land?　　(2)

7 Why is our landscape becoming flatter?　　(3)

8* **Put these types of weathering in order according to how powerful they are in Britain — rain action, freeze–thaw action and onion-skin weathering. Give reasons for your answer**　　**(5)**

3 Glaciated upland landscapes (1)

Location of glaciated uplands

There are four main agents of erosion (wind, waves, rivers and moving ice) and only moving ice does not affect the British Isles today. Yet moving ice has had so much effect on the shape of our landscape that we must study it in detail here.

Our Ice Age ended about 10 000 years ago but the ice was so powerful that it left behind striking landforms that can be clearly seen today. This evidence of glaciation is most obvious when you look at our upland areas.

During the depths of the Ice Age all of Britain except the very far south, was buried under slowly-moving ice-sheets. Only in our highest mountains did glaciers get the opportunity to form and move down the mountainside. When they did this, the mountains never looked the same again because the glaciers eroded them mercilessly. Figure 3.1 shows the main glaciated uplands of the British Isles that were affected in this way.

Processes of glacial erosion

Between a glacier and the rock underneath is a thin film of meltwater. This water sometimes freezes so the glacier becomes attached to the rock. Then, **when the glacier moves forward, it pulls away any loose fragments of rock. This powerful process is called plucking.** It is highly likely that the rock will have loose fragments because it will have been weathered by freeze–thaw action.

Once the glacier has plucked away pieces of rock, they become embedded in the bottom of the glacier and scrape and smooth the rock surface as the glacier moves. This slower process of erosion is called **abrasion**.

Corries

At the start of the Ice Age, snow collected in hollows high up in the mountains and was gradually squeezed into ice. As more and more

Figure 3.1 Glaciated uplands of the British Isles

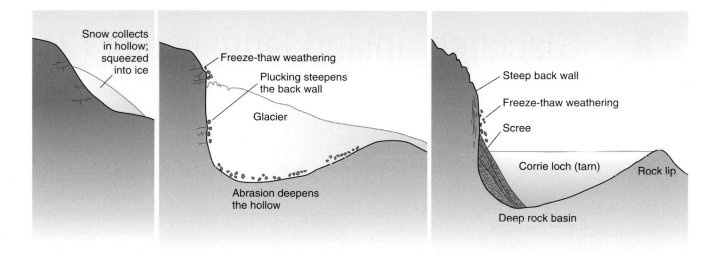

Figure 3.2 Stages in the formation of a corrie

snow built up, it filled the hollow and some of it was then squeezed out and forced down the mountainside. This was the starting point of a glacier (stage 1 in Figure 3.2).

The hollow in which the snow and ice collected was eroded by the ice to form a much deeper, steeper hollow. This is a **corrie**.

As the meltwater under the ice seeped into cracks, **the rock in the hollow was weathered by frost action. Then, when the meltwater froze onto the rock, plucking took place.** The backwall, sides and base of the hollow were eroded very quickly by these two processes. At the edge or lip of the hollow the ice was less thick and so did less plucking. But, by now, there were fragments in the bottom of the glacier and so **the rock at the lip was abraded**, making it smoother (stage 2).

When the ice finally melted at the end of the Ice Age, **corries were sometimes filled with meltwater** and so formed corrie lochs or tarns (stage 3).

QUESTIONS

1 Which is the largest glaciated upland in
   a) Scotland,
   b) England,
   c) Wales? (2)

2 Which part of the British Isles was not glaciated? (1)

3 Describe the process of glacial plucking (3)

4 Describe the process of glacial abrasion (2)

5 Describe and explain, with the aid of diagram(s), how a corrie forms (4)

6* **Look at Figure 3.1. Describe the areas of the British Isles that were**
   a) glaciated
   b) not glaciated (5)

4 Glaciated upland landscapes (2)

Aretes and pyramidal peaks

On many mountains in Britain there is not just one hollow high up near the summit, but several. So, while a glacier was eroding a corrie on one side of a mountain, on the other sides would have other glaciers eroding more corries.

Where two corries formed back to back or side by side, the rock between them was plucked away to form a narrow ridge, shown in Figure 4.1. This steep and narrow ridge is **called an arête**.

Where three or more corries formed back to back the rock between them was plucked and weathered into a sharp point, usually the highest point in the area. This sharp point, shown in Figure 4.1, is **called a pyramidal peak**.

Figure 4.2 This corrie and arête can be found upon Cader Idris, a mountain in Snowdonia, North Wales

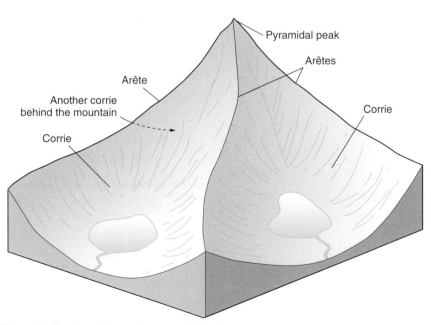

Figure 4.1 Corries, arêtes and a pyramidal peak

U-shaped valleys

When a glacier was squeezed out of a corrie, it moved downhill under gravity. It usually took the steepest route, which, in most cases, was an old river valley. But the glacier was much more powerful than the river that was there before and it was able to completely change the shape and appearance of this valley.

Before the Ice Age **rivers in mountains ran down V-shaped valleys**, but when a glacier rumbled down the same valley **the ice was so thick that it was able to pluck and abrade the valley sides as well as the valley floor. So the V-shaped valley** became steeper and deeper and gradually **took on a more U-shaped appearance** with steep sides and a flat base (see Figure 4.3).

Now that the ice has all gone, these U-shaped valleys have rivers flowing through them again. But **the rivers are too small for these very wide valleys and are called misfit streams**. At the sides of the valley, scree often builds up from all the freeze–thaw weathering that has taken place on the valley sides above.

U-shaped valleys, together with corries and corrie lochs, arêtes and pyramidal peaks, are common sights in the mountains of the British Isles and they are all evidence that we were severely glaciated, not long ago in geological time.

BEFORE THE ICE AGE

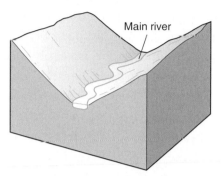

Main river

DURING THE ICE AGE

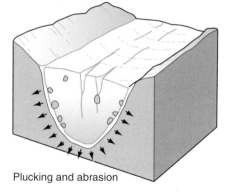

Plucking and abrasion

AFTER THE ICE AGE

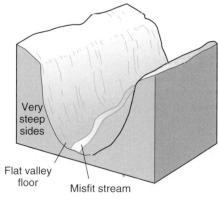

Very steep sides

Flat valley floor

Misfit stream

Figure 4.3 The formation of a U–shaped valley

QUESTIONS

1 Describe the shape of
a) an arête
b) a pyramidal peak   (2)

2 Describe the processes involved in the shaping of arêtes and pyramidal peaks   (4)

3 Explain how a U-shaped valley has formed. You may use diagrams in your answer   (3)

4 What is a 'misfit stream'?

5* **Look at Figure 4.3 Describe how the shape of the valley changes in the three diagrams**   (5)

5 Glaciated upland landscapes (3)

Hanging valleys and truncated spurs

As it moved down the mountainside a glacier was strong enough to keep a straight course. Instead of going around any obstructions, it went over them, quickly eroding them away. **Any spurs of rock that jutted into the valley would be eroded to become truncated spurs.** These became part of the sides of a U-shaped valley.

Along its journey down the valley a glacier would be joined by tributary glaciers. These **smaller tributary glaciers contained much less ice and so were less powerful. They could not erode their valleys as deeply as the main glacier so, where they met, the tributary valley was left 'hanging' above the main valley** (see Figure 5.1). After the Ice Age, when rivers took over, the hanging valley would become a waterfall (Figure 5.2).

Ribbon lakes

At some points along the valley the glaciers were able to erode more deeply than elsewhere. This might have been because the rock there was softer and more easily plucked and abraded. It might have been because the ice became thicker and therefore more powerful. **Wherever the ice did this, it would make a hollow which, after glaciation, would become a lake.** The lake would take on the same shape as the valley in which it was formed, so **it would be long and quite narrow and is usually called a ribbon lake** (see Figure 5.3).

Valley before glaciation

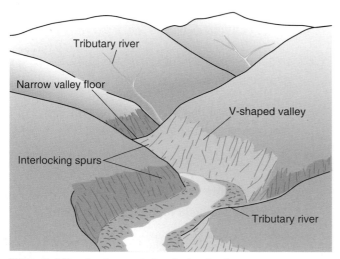

Valley after glaciation

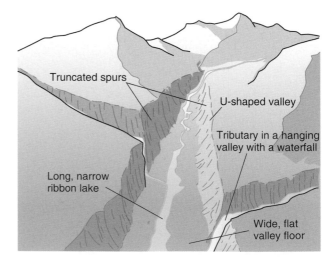

Figure 5.1 The development of a glaciated valley

Figure 5.3 Ribbon lake

When the ice reached warmer regions it melted and dropped all the pieces of rock it had plucked and abraded. This material is called moraine and often forms ridges. Sometimes these **ridges of moraine did not allow the meltwater to escape and so lakes built up behind them**. These moraine-dammed lakes are also examples of ribbon lakes.

Figure 5.2 Hanging valley

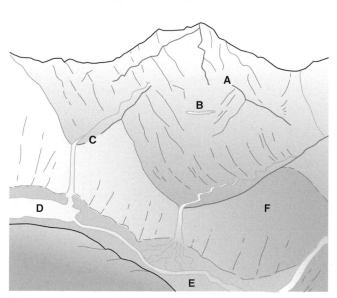

Figure 5.4 Features of glaciated uplands

QUESTIONS

1 Describe the appearance of a hanging valley **(2)**

2 Using the terms 'plucking' and 'abrasion', explain how a hanging valley forms **(4)**

3 Describe two ways in which ribbon lakes form **(6)**

4 Look at Figure 5.4. Six glacial features are shown by the letters A–F. Which letters shows the location of
a) a corrie loch,
b) arête,
c) U-shaped valley,
d) hanging valley,
e) truncated spur,
f) ribbon lake? **(3)**

Look at Figure 5.1. What are the differences between the two diagrams? **(5)**

6 The Lake District—a glaciated upland (1)

The physical geography

The Lake District is England's highest upland area, with mountains rising to over 1000 metres at Scafell Pike. It is also the country's best example of a glaciated upland with much evidence of a variety of glacial features including ribbon lakes, for which the area is famous. Four main rock types make up the area. The hardest are the volcanic rocks, which are examples of igneous rocks, and these form the highest and steepest land, including Scafell Pike. The second hardest rock is slate, a metamorphic rock, of which mountains such as Skiddaw are made. The softest rocks are the grits and limestones, which are sedimentary. They form the lower land at the edge of the Lake District.

As with all of our mountains, the Lake District used to be much, much higher than it is now, but it has been weathered and eroded by many forces, and especially by glaciers. During the Ice Age, glaciers spread out in all directions from the highest part of the Lake District and eroded a trail of features including corries, arêtes, pyramidal peaks, U-shaped valleys and ribbon lakes, the most famous of these being Lake Windermere, the longest lake in England.

Since the Ice Age, rivers have taken over, eroding the land and depositing material elsewhere. Some rivers have deposited so much material into lakes that the lakes are now much shallower and, in time, will fill up. Frost action has also continued from the Ice Age and the bottoms of many of the slopes are strewn with scree material, which has been weathered from the rocks above and fallen down onto gentler slopes. Glaciers have had a huge effect on the scenery of the Lake District and the scenery itself has had a huge effect on the people who live and work here.

Land uses

The land in the Lake District has many uses. Four of the most important ones are described here. Tourism, another important land use, is studied in Chapters 8 and 9.

Hill-sheep farming

There are many different types of farming throughout Britain but very few of them would be profitable in the Lake District. It would be a foolish farmer who tried to grow crops here. **Many of the slopes are too steep for large machinery to be used** and, if the farmer cannot use machines such as combine harvesters, **it is impossible to grow crops**. Being very high up, **the temperatures here are very low** and **the**

Figure 6.1 Striding Edge arête

growing season too short for crops, while the high rainfall means there is **little sunshine to ripen the crops**. The heavy rain also makes **the soil infertile**. Under these conditions not only is arable farming impossible, but so is dairy farming and market gardening. **The only type of farming possible is hill-sheep farming**, although the farmers may be able to keep some cattle on the lower and better land and possibly also grow some grass for making hay and silage.

Forestry

Large plantations of forestry are a common sight in the Lake District, a much more common sight than they are in lowlands. This is not because trees grow better in uplands, but because the land in lowlands is too valuable to be used for forestry. In the Lake District, **because the land is very poor and difficult on which to build, forestry is as profitable and worthwhile as any other activity**. So, over the last 80 years, the Forestry Commission (or Forestry Enterprise, as it is now called) has **planted large areas of coniferous trees**. They have planted coniferous trees (pine, spruce, larch, fir) as these grow better in the cold climate and thin, poor soils. More than 75 000 tonnes of timber are produced from these trees each year.

Industry

Very few factories and offices are attracted to the Lake District. Modern industry prefers to be near its market: **the Lake District has few large towns nearby** and **most of its roads are narrow and slow**. Because very few people live here, a company may also have **difficulty in finding enough workers with the skills** it needs. In addition, there is **a shortage of flat land suitable for building**.

Extractive industry has always been more important here than manufacturing. **Lake District slate** has been used on the roofs of buildings all over the world, including prison buildings in USA, libraries in Denmark and banks in New Zealand. **Granite is also quarried** for use in making roads and limestone is extracted to use in steel works. Over many hundreds of years local rocks were used to build the many miles of drystone walls as well as the walls of the older buildings here. Only ten quarries are still open today.

Water supply

For over 100 years **the lakes of the Lake District have been used to supply fresh water to the people of Manchester**. Although the lakes are 150 kilometres from Manchester, the city has few places nearby that are suitable and **the natural lakes here are cheaper to use than constructing a reservoir**. Being very rainy also means **there is plenty of water available**. At present, Haweswater, Thirlmere, Ullswater and Windermere are used by North-West Water and together they supply 30% of all the water that the region needs.

QUESTIONS

1. Which rock type forms the highest land in the Lake District? Explain why (2)

2. Why is hill-sheep farming so common in the Lake District? (6)

3. Explain why there are many forestry plantations here (2)

4. Which rocks are quarried in the Lake District? (2)

5. What are the advantages of using Lake District lakes as reservoirs? (4)

6* **Describe the landscape and the way the land is used in Figure 6.1.**

7 The Lake District—a glaciated upland (2)

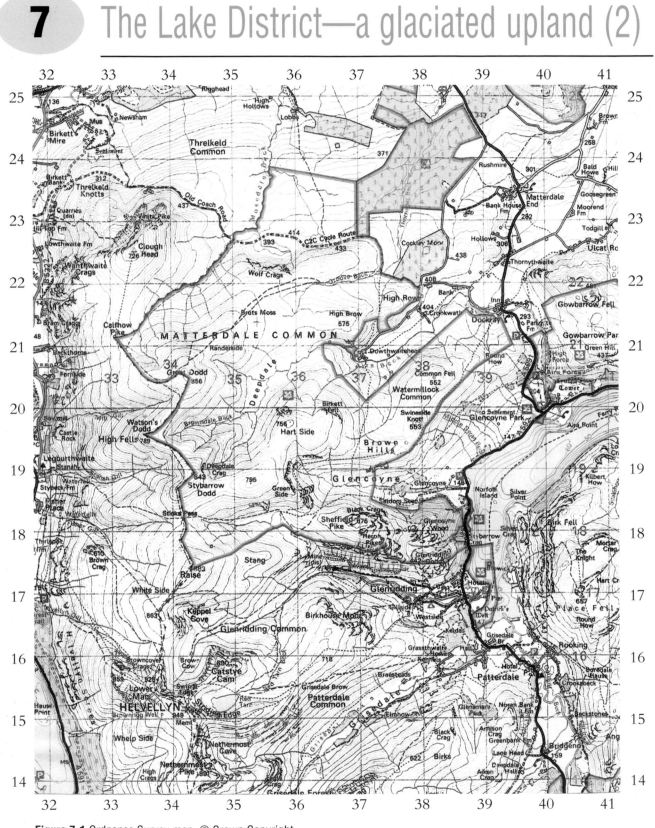

Figure 7.1 Ordnance Survey map. © Crown Copyright

Source: Ordnance Survey Landranges Map number 90 between eastings 32 and 41 and between northings 14 and 25

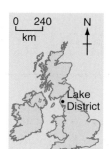

Figure 7.2 Location of Lake District

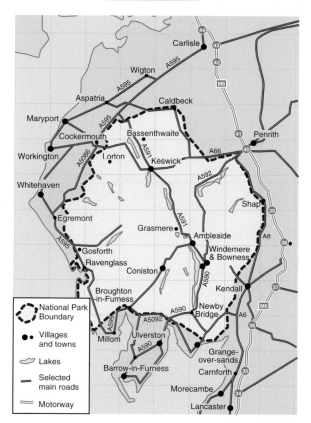

Figure 7.3 Lake District National Park

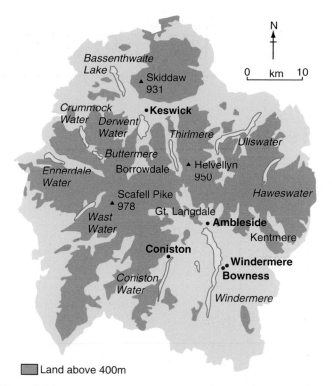

Land above 400m

Figure 7.4 Upland areas in the Lake District

QUESTIONS

Look at the O.S. Map on the opposite page. It shows the area around Helvellyn in the Lake District.

1 Which of these squares has a U-shaped valley:
 a) square 4014 or square 3516?
 b) square 3523 or square 3715? **(2)**

2 What glacial feature is
 a) Red Tarn (3415),
 b) Nethermost Cover (3414),
 c) Striding Edge (3415)? **(3)**

3 Give a grid reference of one ribbon lake on the map extract **(1)**

4 Suggest how the waterfall, Aira Force (square 3920), was formed **(4)**

5 Suggest the type of farming at Bank House Farm (square 3913) **(4)**

6 a) What map evidence is there that this area is popular with tourists? **(4)**
 b) Suggest reasons why tourists visit this area **(6)**

7 What evidence is there of extractive industry on the map extract? **(2)**

8 Describe the distribution of forestry in the area of the map **(3)**

Do you think that many tourists visit the area shown in Figure 7.1? Give reasons for your answer **(5)**

8 The Lake District—a glaciated upland (3)

Tourism statistics

year	visitor days
1974	12 million
1979	12 million
1984	14 million
1989	16 million
1994	22 million
1999	24 million

Table 1 Number of visitors

day visitors	83%
staying visitors	17%

Table 2 Types of visitor

visitors arriving by car	89%
visitors arriving by public transport	11%

Table 3 Transport used by visitors

for active recreation	32%
for sightseeing and recreation	30%
only for sightseeing	25%
just for relaxation	10%
for other reasons	3%

Table 4 Main reason for visiting Lake District

January–March	15%
April–June	30%
July–September	35%
October–December	20%

Table 5 Seasons when visitors come

Attractions of the Lake District

Scenery

For sightseers, the scenery in the Lake District is the most spectacular in England. **There are 16 major lakes**, together with **high peaks** and steep-sided **valleys** with **waterfalls** cascading over their sides. People also come to see the human landscape—the **farmland** with its maze of **drystone walls** and the **pretty villages** made of local stone.

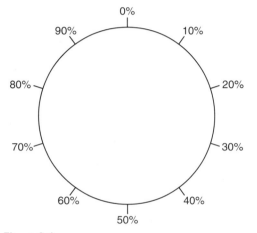

Figure 8.1

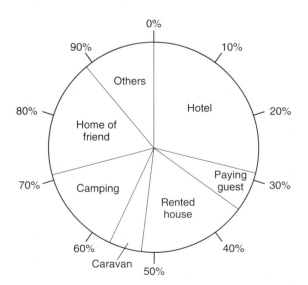

Figure 8.2 Accommodation used by visitors to the Lake District

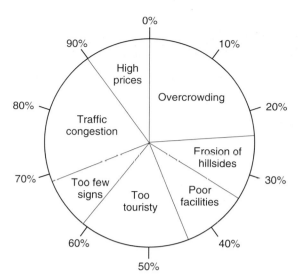
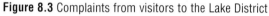

Figure 8.3 Complaints from visitors to the Lake District

People also come here **for recreation**. The lakes provide opportunities for **water sports** of all kinds. The craggy mountains are challenging for skilled **mountaineers** whereas the smoother, gentler hills are popular with other **hill-walkers**. The many rivers here are used for energetic pursuits such as **white-water canoeing**, but can also be used for quieter activities such as fishing.

Historical and cultural features

The peace and beauty of the Lakes have, over the years, attracted many of England's finest poets including William Wordsworth. His home was at Dove Cottage in Grasmere and is open to the public. This area is also the home of Beatrix Potter and there is a Beatrix Potter Exhibition in Bowness and a gallery in Hawkshead. There are many museums, including a pencil museum in Keswick on the site where pencils were once made using local raw materials. Several castles survive, including Muncaster and Sizergh.

Amenities

For any area to attract large numbers of visitors, there must be things for people to do, places for them to stay and fast roads to get them there quickly.

Thanks to the M6 motorway, 24 million people live within a three-hour drive of the Lakes, including the cities of Manchester, Liverpool, Sheffield and Leeds. Once they are there, they can choose from all types of accommodation, including many camp and caravan sites, guest houses, hotels and youth hostels. There are also purpose-built holiday villages and timeshare apartments. There are no large shopping centres or department stores, but the area has many specialist shops and a variety of entertainments, including cinemas and theatres, and a small theme park near Penrith.

QUESTIONS

1 Describe the changes in the number of visitors to the Lake District since 1974 **(2)**

2 Look at Figure 8.2
a) What is the main type of accommodation and how many people use it? **(2)**
b) Which is more popular—camping or caravanning? **(1)**
c) What percentage of visitors stay in rented housing? **(1)**

3 Look at Figure 8.3. Which is a bigger problem, according to visitors:
a) erosion of hillsides or traffic congestion? **(1)**
b) poor facilities or overcrowding? **(1)**

4 Make two copies of the pie-graph in Figure 8.1
a) Complete one pie-graph to show the main reasons for visiting the Lake District, given in Table 4 **(4)**
b) Complete another pie-graph to show the seasons in which visitors come, given in Table 5 **(4)**

6* **The Lake District Tourist Board would like more families with young children to visit the area. Design a poster that could be used to attract them** **(5)**

9 The Lake District—a glaciated upland (4)

The impact of tourism

Land-use conflicts

Of all the land uses in the Lake District, tourism causes the most conflicts. Although it brings many benefits to many people, a lot of the villagers and farmers complain about tourists.

Conflict 1—tourists vs residents

During the summer, and especially on Bank Holidays, **the Lake District roads are congested with slow-moving tourist cars, buses and caravans**. This is inconvenient to the local people and it **increases air pollution**, but is much more serious if emergency vehicles, such as ambulances and fire engines, are delayed. Most tourists head for places such as Bowness and Keswick, where there are shops, cafes and other services, as well as different types of accommodation and places of interest to visit. **These popular spots, called honey-pots, are where traffic congestion is at its worst**. It is difficult even for local people to park their cars in these places.

Because they are so popular with tourists, the centres of honeypots have been taken over by tourist shops and services, meaning that **there are fewer convenience shops** that the residents need. The village of Grasmere, for instance, has 24 shops selling outdoor equipment, antiques and gifts, but only one shop selling bread and milk. What is more, **the goods are sold at higher prices**.

Some of the tourists buy houses in the most attractive parts of the Lake District as second homes. **One out of every six homes here are now holiday homes** or second homes and, in some villages, it can be as much as one-half. But the second-home owners only live there during part of the holidays and at some weekends. So, for most of the time, they do not use the village shop or library or bus service and they do not send their children to the local school. With fewer people using them, **many services close down**. This annoys the local residents as they now have further to travel to do their shopping or take their children to school. It may tempt them to move away, making the problem even worse. Also, once outsiders start buying up houses in attractive villages, their prices rise dramatically. **Many local people will not be able to afford the high house prices** and will be forced to live elsewhere. This breaks up communities and especially reduces the number of families with young children.

Conflict 2—tourists vs tourists

Honeypots in the Lake District include lakes, such as Windermere. They offer many opportunities for recreation and leisure. People can use motor boats or go water skiing. But the people who enjoy these **noisy pursuits conflict with those who are there to escape from the noise of the cities**. These tourists wish to enjoy the views around the lakes and the peace and tranquility. Some wish to windsurf, some fish while others just want to sightsee. None of them want noisy tourists in motor boats creating large waves on the lakes.

A lot of tourist activity can also destroy the scenery that other tourists go to enjoy. It might be walkers eroding hillsides or gaudy camp sites spoiling the views or just the sheer number of people in the honeypots that make it impossible to appreciate the true beauty of the area.

Conflict 3—tourists vs farmers

Many visitors to the Lake District go to explore the hills. Once people leave behind the roads and villages, they need to walk over farmland. When they do this many problems arise. If **the people do not obey the Country Code** they will annoy farmers. Gates left open mean that animals will stray onto roads and be killed or cause accidents. Walkers may leave litter which, as well as looking unsightly, may kill animals if they try and eat it. Visitors may also bring their pet dogs, which can worry sheep and lambs.

When many people walk up a hillside they trample the vegetation until it dies. This creates a 'path' of bare soil which is easily washed away by the frequent rainstorms. Other walkers then avoid this muddy path and walk over the grass next to it, trampling it and killing it, and so the footpath spreads. **This is called footpath erosion.** It is not just an eye-sore. It means the farmer has less grass and soil and so his land is poorer. Footpath erosion is also caused by mountain bikers, horse riders and off-road vehicles.

Increased employment and wealth

Over 20 000 people are employed in tourism in the Lake District, although **some of the jobs are for only part of the year.** The main types of employment are shown in Table 1. **Farmers also benefit from tourists because they can rent out their land** for camping, caravanning or for recreation such as scrambling. They can also sell their produce directly to the visitors. **House owners gain if they sell their houses to second-home owners at high prices.** And the richer second-home owners are more **likely to afford to employ tradespeople to improve their new homes.**

The extra jobs and money brought by tourism is spread throughout the community by the multiplier effect. People who now have a greater income will spend more money (e.g. in restaurants, travel agents, furniture shops). This means that other people become richer and so employ more workers, who spend more money in local shops, etc. In this way, far more than the 20,000 people directly employed benefit from tourism in the Lake District.

Tourism jobs in the Lake District	
Every £100 000 spent by tourists creates the following number of jobs:	
in accommodation	61 jobs
in camping and caravanning	8
in restaurants/pubs	12
in shops	3
in visitor attractions	13
total	97

Table 1

QUESTIONS

1. What is a 'honeypot'? (1)

2. Describe the problems tourists cause at honeypots in the Lake District (6)

3. Explain how tourists increase the amount of air pollution, noise pollution, water pollution and visual pollution (4)

4. Explain how tourists cause footpath erosion (3)

5. Describe one other way in which tourists upset farmers (2)

6. Explain how local people benefit from tourists (4)

7. What is meant by the 'multiplier effect'? (3)

- Do you think it would be better for the local people if fewer tourists visited the Lake District? Give reasons for your answer (5)

10 The Lake District—a glaciated upland (5)

Many organizations are involved in trying to protect and conserve the Lake District's beauty and to sort out the problems between the different land users. These organizations fall into two categories: public or official bodies, and voluntary bodies.

The National Park Authority — an official body

The largest official body trying to manage conflicts is the government. It has created 11 National Parks (see Figure 10.2) in those areas of the country where there are the greatest number of land use conflicts, including the Lake District.

Figure 10.1 National Park logo

A National Park is a type of conservation area. Before any new tourist development can take place, the plans have to be approved by the Planning Board and **they refuse planning permission for any scheme that will cause conflicts between tourists and local people**, such as increasing traffic congestion or spoiling views.

The National Park Authority also tries to reduce traffic congestion by a variety of means. It has brought in one-way systems, for example in Ambleside. In towns such as Keswick some of the streets in the centre have been pedestrianized. In others, such as Grasmere, no street parking is allowed at all, but large car parks are provided at the edge of the village. The National Park Authority also tries to reduce the number of cars and coaches in honeypots by advertising and signposting other attractions in different areas. It

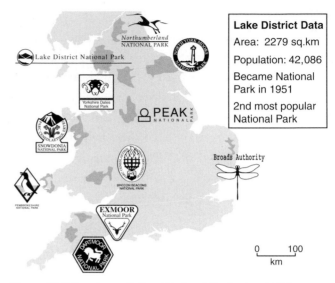

Lake District Data
Area: 2279 sq.km
Population: 42,086
Became National Park in 1951
2nd most popular National Park

Figure 10.2 The eleven National Parks of England and Wales

does this at its information centres, where it is also possible to educate and inform the tourists so they are less likely to upset others.

The National Park Authority employs rangers whose job it is to spot and, if possible, prevent problems between different types of tourists. **There is also zoning of tourist activities**. Only some activities are allowed in different areas or zones so there are fewer conflicts there.

The National Park Authority also has some control over new housing. **It can insist that new houses are sold to local people.** For example in Rosthwaite, a popular second-home village, a row of affordable terraced houses was built for local people only, whereas at School Knott near Bowness, thirty new houses have been constructed that are only available for local people to rent. **The National Park Authority also encourages timeshare developments,** which provide alternative places to buy and so might reduce the demand for other village properties.

The role of voluntary bodies — The National Trust

Figure 10.3
National Trust logo

The National Trust is a voluntary conservation body set up in 1895. It gets its money from donations from the general public. It conserves the beauty of the Lake District **by buying land and buildings and then managing them itself.** At present it owns one-quarter of all the land in the Lake District, including 91 farms. In this way it can ensure that at least the land it owns is protected and not used in a way that will upset others. For example, it reduces conflicts between tourists and farmers by **maintaining drystone walls and important wildlife habitats** on its land; it also **reduces footpath erosion** caused by hill-walkers. Volunteers have laid large blocks of hard-wearing stone on eroded hillsides to provide a good walking surface. This is called stone-pitching. The stone is of local rock so that it blends in with the landscape. The volunteers also fence off the worst affected areas to allow the vegetation time to recover, and dig out drainage channels down the hillside so the rain does not run down the eroded footpaths, taking soil with it.

Sustainable tourism

We have seen that tourism brings many benefits to the people of the Lake District, but that it brings problems as well. There is a danger that, in the future, these problems may worsen and the area will be much less attractive to visitors.

The National Park Authority and the National Trust both want people to visit the Lake District but they also want to ensure that, while they are there, they do not damage the environment for future tourists. This is called **sustainable tourism**. Table 1 shows some examples of sustainable tourism used in the Lake District.

Examples of sustainable tourism
1. careful planning of tourist facilities
2. zoning of tourist activities
3. limiting the numbers of cars in villages
4. protecting the hillsides from footpath erosion
5. educating visitors

Table 1

QUESTIONS

1 Name one official body and one voluntary body that looks after the Lake District **(1)**

2 How does the Lake District National Park Authority
   a) prevent new developments from taking place,
   b) reduce the traffic congestion caused by tourists,
   c) reduce the problem of noise pollution caused by tourists? **(6)**

3 By what means does the National Trust protect the Lake District countryside? **(2)**

4 Describe the ways in which the National Trust reduces footpath erosion on hillsides **(3)**

5 What is 'sustainable tourism'? **(2)**

6 Choose two of the examples in Table 1. Explain how each one promotes sustainable tourism **(4)**

7* **Some people think that cars should be banned from parts of the Lake District. Do you think this is a good idea? Give reasons for your answer** **(5)**

11 Upland limestone landscapes (1)

Location of limestone uplands

We found out in Chapter 2 that some rocks are eroded more quickly than others. Similarly, some rocks are worn down by weathering more quickly than others. One such rock is limestone.

There are many types of limestone in the British Isles, but the one that forms the highest upland areas is carboniferous limestone. It formed about 350 million years ago and is found mostly in northern England (e.g. Yorkshire Dales, Peak District) and in Ireland (e.g. The Central Plain). These areas are shown in Figure 11.1.

Processes affecting limestone landscapes

A lump of limestone is very hard—hard enough to be used as a building material. The Pyramids of Egypt are made of limestone and they have lasted a few thousand years and will look good for another few thousand years more. But **limestone is made of calcium carbonate and calcium carbonate suffers from chemical weathering.**

When raindrops fall through the atmosphere they absorb carbon dioxide from the air, which makes rainwater a very weak acid. When it reaches rocks such as limestone, the acid in the raindrops starts to dissolve the calcium carbonate and then removes it in solution. This is **the process of solution and is a type of chemical weathering.** This **is the main way in which limestone begins to wear down**. Just like the processes of erosion, the processes of weathering are very slow. Limestone dissolves at a rate of about 1 cm in every 250 years.

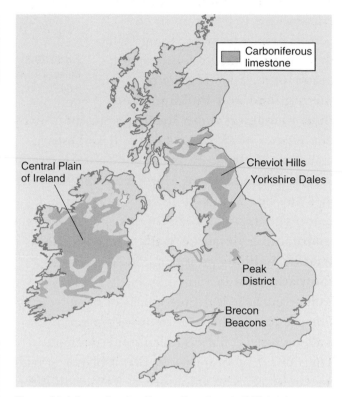

Figure 11.1 Area of carboniferous limestone in British Isles

Limestone pavement

Limestone suffers badly from chemical weathering—not just because it is made up of calcium carbonate, but also because it has lots of cracks in it. As is shown in Figure 11.2, **limestone has horizontal cracks called bedding planes and vertical cracks called joints.** This means that very little water is found on the surface. Instead, it seeps through the cracks and into the rock. **Such a rock is said to be permeable.**

In areas of upland limestone in Britain, the soil has long since been removed by ice-sheets, so the surface is just dry, bare rock. When rain falls onto this surface, it seeps into the many joints. As it does so, it dissolves the rock on either side by

(i) Before weathering

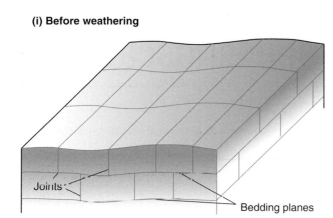

Joints

Bedding planes

(ii) After weathering Clints Grikes

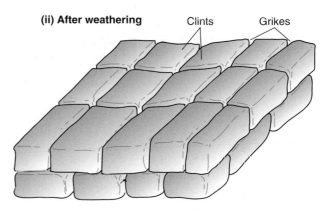

Figure 11.2 The formation of a limestone pavement

Figure 11.3

chemical weathering, making the joint wider and wider. This happens over the whole surface of the limestone so that, eventually, **the surface is broken up into a series of rectangular blocks separated by wide, deep cracks.** This feature is called a **limestone pavement** (see **Figure 11.2**) and the blocks are called **clints** and the enlarged cracks are **called grikes.**

Swallow holes and pot-holes

Because limestone is a permeable rock, there are few surface streams. Streams that flow onto

limestone quickly fall into one of the many enlarged joints on the surface and disappear underground. **Where the river goes underground is called a swallow hole.** It is usually seen as a small hollow or depression, below which is a deep, wide, vertical crack (see Figure 11.3), although sometimes this crack can be seen at the surface. **This feature is sometimes called a pot-hole.**

QUESTIONS

1 Name three upland limestone areas in the British Isles   (2)

2 What type of limestone forms the highest uplands?   (1)

3 Describe fully how limestone is worn away by the process of solution   (4)

4 What are 'joints' and 'bedding planes' and how are they different?   (2)

5 What is meant by a permeable rock?   (1)

6 Describe a limestone pavement landscape   (3)

7 Explain how limestone pavement forms   (3)

8 Describe the appearance of a swallow hole (2)

List as many facts about limestone as you can find on these two pages   (5)

12 Upland limestone landscapes (2)

Caverns

The water that enters the limestone through a swallow hole makes its way down through the rock. It flows along the many joints and bedding planes, dissolving the limestone as it goes. **Caverns form where some of the underground limestone is dissolved more quickly than the rock around it. This happens in places where the rock has many joints and bedding planes close together** (Figure 12.1A). These cracks allow through lots of water, which dissolves away the rock completely and a cavern forms (Figure 12.1B).

Stalactites and stalagmites

The water that drips down into the caverns is laced with calcium carbonate that has dissolved on its passage through the rock. The water drips from the cavern roof very slowly so that some of it evaporates. **When it evaporates it leaves behind the calcium carbonate**, which is deposited on the cavern roof. When calcium carbonate is deposited it is called dripstone. The water continues to drip, evaporating as it does so, and **the deposits build up to form fingers of dripstone that grow downwards into the cavern** (see Figure 12.1B). They are called **stalactites** and grow by only a few millimetres a year. They grow slowly, partly because the water cannot hold much dissolved limestone and partly because the caverns are cool so only a little evaporation takes place.

A. Water passes through cracks in limestone

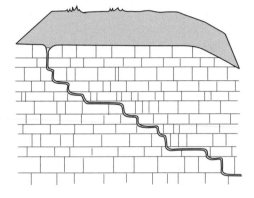

B. Caves form where joints and bedding planes are close together

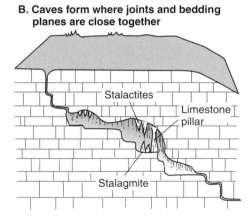

Figure 12.1 The formation of caves

Some of the water drips onto the cavern floor where it also may evaporate. It leaves behind calcium carbonate here as well, which is deposited as dripstone on the cavern floor. **As more water drips down, more is deposited, forming fingers of dripstone that grow upwards from the cavern floor.** They are **called stalagmites** (see Figure 12.1B) and grow just as slowly as stalactites. Sometimes stalagmites and stalactites join together to form limestone pillars (see Figure 12.1B).

Gorges

Water passes through limestone all the time, dissolving the rock as it does so. Over time, therefore, the joints and bedding planes become wider and wider and the caverns become bigger and bigger. **As a cavern becomes higher, there is less and less rock left above it. This rock becomes unstable** because there is nothing underneath to support it. It starts to crack even more and **eventually collapses**, the pieces of limestone falling to the floor of the cavern. This forms a very deep, steep-sided gash or valley, called a **gorge**. Possibly the most well-known gorge formed in this way is Cheddar Gorge in the Mendip Hills of England.

Intermittent drainage

Streams flowing onto limestone rock disappear down swallow holes. They flow downwards, often in a zig-zag route, along bedding planes and down joints until they reach the water table. The water table is the furthest point that water can reach and is usually an impermeable rock. At this point the underground streams have to flow along the water table until they reach the surface

again at a much lower level. Where underground water comes to the surface is called a **spring**.

After heavy rain the water table sometimes rises (see Figure 12.2). At these times the underground streams come to the surface higher up and flow down the slope into the river. **Because these streams do not flow over the surface all the time they are called intermittent streams and the area is said to have an intermittent drainage.**

Because limestone has so few surface streams the area is said to have a low drainage density.

QUESTIONS

1 Explain how underground caverns form in limestone areas **(3)**

2 What is the difference between a stalactite and a stalagmite? **(1)**

3 Explain how stalactites have formed **(4)**

4 Explain how a limestone gorge forms **(4)**

5 What is meant by the following terms: **a)** water table, **b)** spring, **c)** intermittent drainage, **d)** a low drainage density? **(4)**

6* **Look at Figure 12.3 which shows an upland limestone area opposite. Match the numbers on the diagram to the following features: a) swallow hole, b) cavern, c) stalactite, d) stalagmite, e) gorge, f) spring** **(6)**

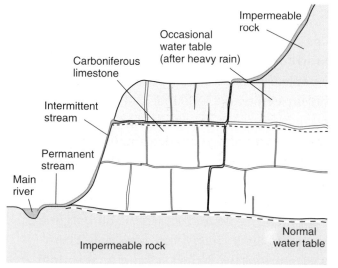

Figure 12.3 An upland limestone landscape

Figure 12.2

13 The Yorkshire Dales—a limestone upland (1)

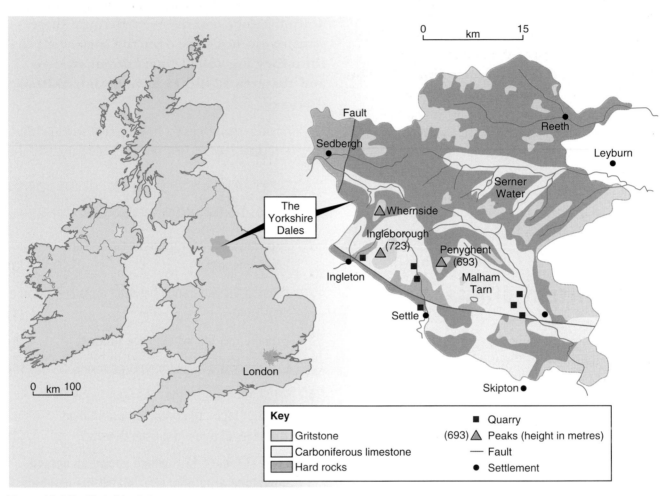

Figure 13.1 The Yorkshire Dales

The physical geography

Running down the centre of northern England, and separating the counties of Yorkshire and Lancashire, are the Pennine Hills. Rising to 900 metres, they are just a little lower than those in the Lake District and a lot less dramatic. On their eastern side the hills are cut through by a series of dales or valleys, which have given the area its name.

The Dales are made chiefly of two hard rocks: carboniferous limestone and millstone grit. It has the largest area of upland limestone in mainland Britain and the best limestone scenery. At the end of the Ice Age glaciers stripped the land here of its soil, leaving a bare limestone surface complete with limestone pavement and swallow holes (see Figures 13.2 and 13.4) and cut through by deep gorges such as Gordale Scar (see Figure 13.3). Water plunges underground through pot-holes into huge caverns, such as Battlefield Cavern (see Figure 13.5), which is over 100 metres long and 30 metres high and is decorated with orange-coloured stalactites and stalagmites, some of which are 100 000 years old. The underground rivers then re-emerge as springs at the edge of the limestone.

Figure 13.2 Gaping Gill

Figure 13.3 Gordale Scar

Figure 13.4 Malham Moor

Figure 13.5 Battlefield Cavern

QUESTIONS

1 What are the two main rock types in the Yorkshire Dales? **(1)**

2 What limestone features are **a)** Gordale Scar, **b)** Gaping Gill, **c)** Malham Moor? **(3)**

3 What limestone features are found in Battlefield Cavern? **(1)**

Describe the landscape shown in Figures 13.3 and 13.4 **(5)**

14   The Yorkshire Dales—a limestone upland (2)

Land uses

Upland areas in Britain are not the easiest areas from which to make a living. They all present similar problems to people and so the land uses found in one upland area are very similar to those found in another. Land uses in the Yorkshire Dales are much the same as those in the Lake District (described in Chapter 3) but the limestone rock does present additional problems to people, while providing other opportunities as well.

land use	Lake District	Yorkshire Dales
farming	30%	41%
moorland	54%	56%
forestry	11%	2%
reservoirs	4%	0.1%
quarrying	<0.1%	0.1%
military training	0.2%	0.4%

Figure 14.1 Land use in the Yorkshire Dales and Lake District

A typical Dales farm

livestock: 700 sheep, including 600 lambing ewes producing about 600 lambs a year +25 suckler cattle producing 20 calves a year
farm area: 300 hectares
area of meadow: 20 hectares
area of pasture: 80 hectares
area of moorland/rough grazing: 200 hectares
labour: farmer and son

Figure 14.2

Hill-sheep farming

The climatic problems in the Dales are the same as those in the Lake District. **It is cold, the growing season is short and it is wet and cloudy with little sunshine.** The slopes are not as steep as those in the Lake District but **the soil is much thinner**, being non-existent in areas of limestone pavement. Conditions are made worse by the fact that limestone is permeable so there is **very little water near the surface** that plant roots can reach. All in all, **it is an area completely unsuited to growing crops**, where grass does not generally grow well enough for cattle and where sheep farming is the only possible enterprise. **The valleys here have deeper, more fertile soil and are a little warmer and drier**, so they are often grazed by cattle, or the grass is used for making hay and silage.

Quarrying

Limestone is a very useful rock. Because of its many vertical and horizontal cracks, it **is easy to shape and use for building**. Many of the buildings in the Dales are made of the local limestone. **It is also needed in steelworks, in chemical industries and for making cement.** And the limestone in the Dales is the purest in the country, which makes it even more popular. **The gritstone is also quarried**, and its chief use is in the construction industry. **The biggest single use for both the limestone and the gritstone is in making aggregate, for road surfaces.**

As a result there are eight quarries here employing over a thousand people and bringing in around £6 million to the local economy. Because it is the most important raw material for making cement, **some cement works have also located here.** But **not many other manufacturing industries choose to come to the Dales,** as it is a little too remote from their raw materials and markets, and there are few workers.

Tourism

Thanks to soap operas, such as *Emmerdale*, and serials such as *All Creatures Great And Small*, most people will have seen the special scenery of the Yorkshire Dales on TV. It helps to explain why 8 million visits are made to this area by tourists each year, who spend over £50 million pounds while they are there and so create the equivalent of 1000 full-time jobs. **People are attracted by the interesting and unusual scenery** such as limestone pavements, caverns, gorges and stalactites and stalagmites (Figures 14.3 and 14.4).

As well as looking at the natural and human attractions, there are more energetic activities to do. **Hill-walking is popular** (Figure 14.5). The longest Long Distance Footpath in Britain, the Pennine Way, runs through the Dales and there are over 2000 km of public rights of way here. In addition, the Dales is one of the few areas in the country in which enthusiasts can go pot-holing.

Figure 14.3

Figure 14.4

Figure 14.5

Other land uses

Forestry is less important in the Dales than in any other National Park. This is chiefly because the limestone is so permeable that there is not enough water near the surface for trees to grow well. The only forest plantations are on the more impermeable millstone grit.

For the same reason very little of the Dales is used for reservoirs, although there are many cities nearby needing water. It is much more expensive to build a reservoir on permeable rock.

Military training is a minor land use here. The Ministry of Defence mostly uses uplands because (a) they make suitable training areas, (b) they are not taking over valuable farmland or building land, and (c) there are fewer people in uplands who will be disturbed by their activities.

QUESTIONS

1 Why is it difficult to grow crops in the Yorkshire Dales? **(4)**

2 **a)** How many quarries are there in the Dales? **b)** How many people work there? **c)** Which two rocks are quarried? **d)** What is their main use? **e)** Which manufacturing industry in the Dales uses limestone as a raw material? **(3)**

3 Which limestone features in the Dales are tourist attractions? **(2)**

4 What activities are there for visitors to do in the Dales? **(1)**

5 Explain why forestry is not a major land use in the Dales **(2)**

6 Why is military training common in uplands? **(3)**

7* **Look at Figure 14.1. Compare the land uses in the Yorkshire Dales and Lake District** **(5)**

15 The Yorkshire Dales—a limestone upland (3)

Land use conflicts

As in other areas, the different land users in the Yorkshire Dales do not always get on with each other. Tourism causes similar problems to those found in the Lake District. Another major conflict is quarrying (Figure 15.1). It brings lots of benefits to local people and to industries all over the country but it also manages to annoy many other land users.

Conflict 1—quarrying vs tourists

Many tourists go to the Yorkshire Dales to escape the hustle and bustle of city life and enjoy the peace, quiet and beauty of its countryside. But the peace and quiet can be rudely interrupted by the noise of blasting from large quarries (**noise pollution**). Also, the beauty of the scenery is interrupted by the sight of large, white holes in the ground (**visual pollution**). The blasting causes a lot of dust (**air pollution**), which settles on the land around, making everything an unnatural white colour (**visual pollution**). This blasting can also affect the caverns underneath and make stalactites and stalagmites unstable. All these things annoy tourists to the extent that some may not return to the Dales. And if the **number of tourists decreases**, the **number of tourist jobs decreases** and the amount of money spent in local shops, restaurants and hotels decreases and so all the shop- and restaurant- and hotel-owners suffer as well.

Conflict 2—quarrying vs local residents

Although quarries provide 7% of all the jobs in the Dales, many residents are upset by their activities. As well as the pollution they cause directly, they use fleets of lorries to take away the quarried rocks. These slow-moving lorries cause traffic congestion on the narrow roads, make the roads more dangerous, as well as increasing the air pollution. On busy days, when there are tourist vehicles as well as lorries on the roads, air pollution in the Yorkshire Dales is worse than in central London.

As the demand for limestone in the steel industry has declined, many quarries have closed. But, disused quarries are still an eyesore to everyone. They are deep and steep and so can be dangerous, especially to children playing nearby.

Conflict 3—quarrying vs farmers

The people most affected by quarrying are farmers because they live next door to the quarries. When dust settles on fields, **the crops do not grow as well** because the sunlight cannot get through to them. The noise from **blasting can frighten the farmers' animals,** especially sheep that are ready to lamb. And the dust and waste from the quarries wash into nearby streams making them unsuitable for farm animals to drink from (**water pollution**). And farmers, like everyone else, will be inconvenienced by the extra number of lorries clogging up the country roads.

Figure 15.1 Horton Quarry

Figure 15.2 © Crown Copyright

Source: Ordnance Survey Outdoor Leisure Map no 2 'Yorkshire Dales – Southern – Western Areas' between eastings 75 and 80, and northings 71 and 75

QUESTIONS

1 Describe the types of pollution caused by quarrying **(3)**

2 Apart from pollution, describe one other way in which quarrying upsets tourists **(2)**

3 Describe the effects on local people if the number of tourists falls **(3)**

4 In what ways do quarry lorries upset local people? **(3)**

5 Describe two ways in which quarrying conflicts with farmers **(4)**

Do you think more quarries should be allowed in the Yorkshire Dales? Give reasons for your answer. (5)

16 The Yorkshire Dales: a limestone upland (4)

Management of the Yorkshire Dales

Several organizations in the Dales are concerned with reducing the conflicts between land users, and especially the problems caused by quarrying.

The National Park Authority—an official body

Figure 16.1 Yorkshire Dales National Park logo

The Yorkshire Dales became a National Park in 1954 so **its Planning Board can refuse planning permission for new quarries**, but they do not have the power to close down ones that are already in use. Instead, in extreme cases, **they can buy the land from the quarry company** to prevent them spoiling the landscape any more. They did this in Ribblesdale to protect a valuable area of limestone pavement. The Planning Board can also **insist that companies screen their quarries** with fast-growing trees in order to reduce the visual, air and noise pollution. **Companies must also restore the quarries** after they have finished using them. Either the hole must be infilled and trees or grass planted or it must be turned into a lake and then landscaped for people to enjoy.

The role of voluntary bodies—the Yorkshire Dales Society

Figure 16.2 Yorkshire Dales Society logo

The Yorkshire Dales Society is an educational charity that promotes the conservation of the Dales landscape and the ways of life of its people. **It brings problems such as quarrying to people's attention**. It does this through its magazines, through walks it organizes and lectures it gives. **It makes recommendations**. For example, it recommends that more rocks be transported away from quarries by rail. At present only 7% is taken away by rail, but one quarry company has recently agreed to send more rocks away by rail. This should reduce the number of heavy lorry journeys each year by thousands. **It also informs quarry companies of public opinion**, which has led to them agreeing not to open any disused quarries nor to expand existing ones. **It promotes sustainable development**. For example, it encourages alternative jobs to quarrying in new industries that do not cause environmental damage.

Figure 16.3 Sketch of limestone pavement

Figure 16.4 Fieldsketch of Yorkshire Dales

Figure 16.5

QUESTIONS

1 How does the Yorkshire Dales National Park Authority reduce conflicts caused by quarrying? **(4)**

2 Name one voluntary body at work in the Dales and describe how it also tries to reduce conflicts **(4)**

OS MAP QUESTIONS: Look at the OS Map of the Yorkshire Dales on page 31

3 757714; 758735; 785723; 766734; 794738 Match the five grid references above to the following landforms: spring, swallow hole, cave, gorge, limestone pavements **(4)**

4 Give a grid reference of another example on the map of **a)** a cave, **b)** limestone pavement, **c)** a spring, **d)** a swallow hole **(4)**

5 What is the most likely type of farming at Crummack (GR 773714)? Give reasons for your answer **(4)**

6 Suggest reasons why so few people live in the area of this map **(6)**

7 Look at the OS Map and Figure 15.1 on p. 30 Describe the arguments against the expansion of Horton Quarry (GR 7972) **(4)**

8* Look at the OS Map and Figure 15.1. Suggest what should be done with Horton Quarry (GR 7972) when it is no longer needed as a quarry **(5)**

Techniques Questions

9 Figure 16.3 is a sketch of a limestone landscape. Draw this sketch and annotate (label) it to show the main landforms. **(3)**

10 Figure 14.4 on page 29 shows underground limestone features. (a) Draw a fieldsketch of Figure 14.4, (b) Label the sketch to show the main limestone features **(3)**

11 Figure 16.4 is a fieldsketch of the area shown in Figure 14.3 on p. 29. Draw this sketch and annotate it to show the main land uses. **(4)**

12 Figure 16.5 is a sketch of Figure 14.5 on p. 29. Draw this sketch and annotate it to show the reasons why Malham and the surrounding area is popular with tourists. **(4)**

17   Coastal landscapes (1)

Introduction

In the Middle Ages the city of Dunwich in Suffolk was one of the biggest ports in the region. It had a large fleet of fishing boats and two Members of Parliament. Today, most of the city lies under the sea and only a few hundred people live there.

Further north, along the Yorkshire coast of Holderness, over thirty villages have been lost to the sea since Roman times and waves are still eroding back the coast at a rate of 2 metres a year.

These two areas provide graphic evidence that the sea, just like rivers and moving ice, is a very powerful agent of erosion. It erodes a variety of landforms and, like all agents of erosion, it also transports the material it erodes and makes new landforms where it deposits this material.

Coastal processes

Processes of coastal erosion

Waves that have a lot of energy are able to erode the land at the coast. Figure 18.1 on page 36 shows those areas in the British Isles that suffer the most erosion.

Waves erode the land by three main processes:

★ **hydraulic action – this is the sheer power of the waves crashing against the cliffs**, compressing the air in its cracks, which make the cracks wider and longer until pieces of rock break off (see Figure 17.2)

★ **corrasion – this is when the sand, shingle and pebbles that waves are carrying are hurled against the cliff**, causing pieces of rock to eventually break off

★ **solution** – this occurs as salt and other **chemicals in the sea-water slowly dissolve minerals in the rocks**, causing them to break up.

The rocks that break off the cliff and lie on the beach are then picked up by other waves and used to corrade the cliff even more. The rocks

Figure 17.1 Cliff erosion

Figure 17.2 The hydraulic action of waves

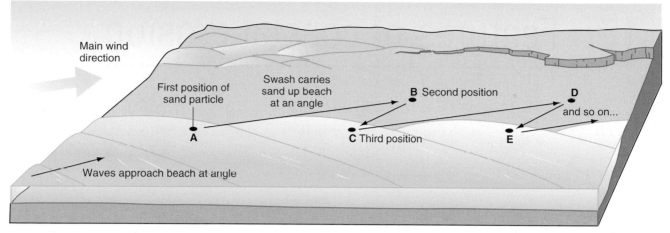

Figure 17.3 The process of longshore drift

themselves are broken up into smaller and more rounded pieces.

Processes of coastal deposition

The coastline of the British Isles is being worn away, but this does not mean that our land will eventually disappear. The material that waves erode is deposited and it builds up land elsewhere. Figure 18.1 on page 36 shows the main areas in the British Isles where coastal deposition is taking place.

Waves deposit in areas where they have very little energy and cannot transport all the material they are carrying. **They deposit the largest particles first, so the material is sorted** according to size.

Processes of coastal transportation

When a wave breaks near the shore and washes up the beach, it is called swash. When it runs back down the beach into the sea, it is called backwash.

Along many of our coastlines the swash travels up the beach at an angle (see Figure 17.3). As it has some energy it picks up sand and shingle (point A in Figure 17.3) and takes them up the beach (point B). But it does not go back along the same route. Instead, the backwash returns to the sea down the steepest slope. The backwash carries

the sand and shingle with it, which it deposits where it loses energy (point C).

The swash from the next wave then picks up the same particles and takes them up the beach at an angle (to point D) and the backwash returns them down the steepest slope (to point E).

In this zig-zag way sand and shingle are transported along a beach. The process is called longshore drift and it takes place in the direction of the prevailing winds.

QUESTIONS

1 Describe the three processes by which waves erode (3)

2 Waves deposit 'sorted' particles. What does this mean? (1)

3 What is the difference between swash and backwash? (1)

4 What name is given to the movement of sand along a beach? (1)

5 Using a diagram(s), explain how material is moved along a beach (6)

Draw a diagram of waves crashing against a cliff and label it to show how waves erode (5)

18 Features of coastal erosion

Cliffs

In places where high land reaches the sea, cliffs form. Here they are attacked by waves that are constantly crashing against them and eroding them by hydraulic action, corrasion and solution. **The waves mostly attack the base of the cliff, which then gets worn away fastest, so that a wave-cut notch begins to form** (see stage 1, Figure 18.2). As the waves continue to pound away at the foot of the cliff, **the wave-cut notch becomes wider and deeper** (stage 2) until the rock above begins to crack. In time, pieces of rock fall off and **then the whole cliff above it collapses into the sea** (stage 3). This process is speeded up by the weathering of the

cliffs, which will be taking place at the same time. The cliff has now retreated and the process starts again with waves eroding another wave-cut

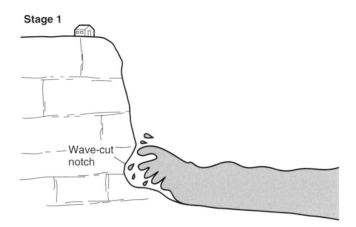

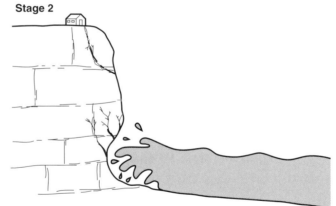

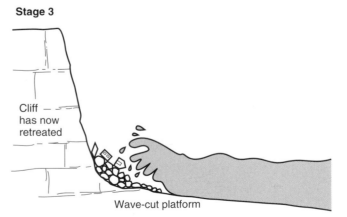

Figure 18.2 Stages in cliff erosion

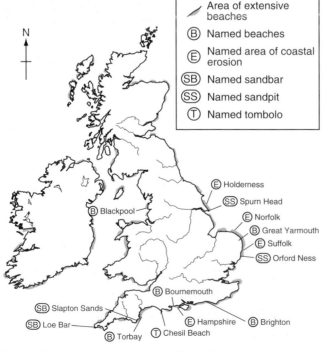

Figure 18.1 Coastal features around the British Isles

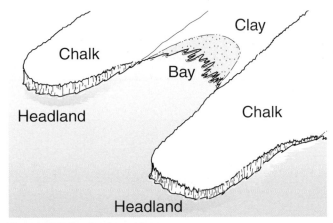

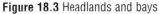

Figure 18.3 Headlands and bays

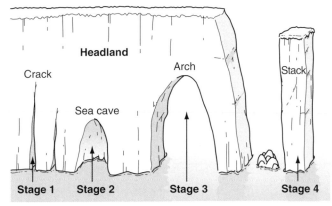

Figure 18.4 Stages in the erosion of a headland

notch in the new cliff face. As the cliffs erode back, **a gently-sloping rock surface is left** in front of them, **called a wave-cut platform.**

Headlands and bays

Like other agents of erosion, waves can erode soft rocks more quickly than they can erode hard or resistant ones. Along coastlines where the cliffs are made of different rock types, the **softer rocks are eroded back quickly to form bays** (such as clay in Figure 18.3). **The harder rocks are eroded more slowly** and left jutting into the sea. **They form headlands** (such as chalk in Figure 18.3).

Caves, arches and stacks

Once headlands and bays form, the headlands then receive the full fury of the waves. **The waves** pound against all sides and **erode first the weakest parts of the headland.** These are the places where the cliffs have cracks. By corrasion and hydraulic action **the waves make the cracks wider** (stage 1, Figure 18.4). In time, this weak area will be eroded more and more until a cave forms (stage 2). The waves now batter away at the sides of the cave and the back of the cave until they cut through to the other side of the headland and **the cave becomes an arch** (stage 3). The rock around the bottom of the arch is now attacked by waves so that it becomes wider. Meanwhile, the rock above the arch becomes more unstable. Cracks appear and, in time, **the rock above collapses. This leaves a pillar of rock separated from the headland, called a stack** (stage 4), which will itself eventually collapse.

The waves continue to erode in this way, widening cracks, forming caves, making arches and stacks, until the headland is completely worn back and the coastline becomes straight again.

QUESTIONS

1 Using a diagram(s), explain how cliffs are worn back　　**(4)**

2 Why do headlands and bays form along coastlines?　　**(2)**

3 Explain how caves form in headlands　　**(2)**

4 What is the difference between a cave and an arch?　　**(1)**

5 Explain how an arch becomes a stack　　**(2)**

6* **Look at Figure 18.4. Draw a sketch to show what this headland will look like in the future. Label the sketch to show what is taking place**　　**(5)**

19 Features of coastal deposition

Beaches

Beaches form where the waves have little energy so that they deposit the mud, silt, sand and shingle that they have been carrying. The largest beaches are usually found in bays, where the waves are generally weak. Beaches are made up of rock fragments that have been eroded from cliffs and then broken up into smaller pieces and rounded off. Sometimes they have been carried along the coastline by longshore drift.

A typical beach has sorted deposits—the largest particles are found at the back of the beach and the smallest ones next to the sea.

The swash from waves carries particles of all sizes up a beach. **When the backwash returns to the sea it loses energy** travelling down a gentle slope. As it loses energy, **it deposits the largest particles** first (see Figure 19.1). The mud and silt are smaller and can be carried much nearer to the sea before they are dropped. If a beach is very steep the backwash will have more energy and so might only deposit shingle before reaching the sea. On more gently sloping beaches, where the backwash has little energy, sand, silt and mud may be deposited as well as shingle.

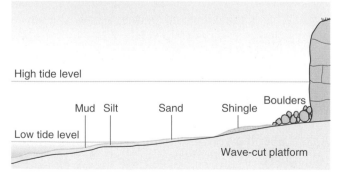

Figure 19.1 Section across a beach

Figure 19.2 A Tombolo

Sandspits, sandbars and tombolos

Longshore drift is the process by which material is carried along a beach. It is responsible for the formation of several coastal features.

When the coastline changes direction, swash will continue to pick up sand (point A in stage 1 of Figure 19.3) **and deposit it in open water** as it runs out of energy (point B). In time, **it deposits enough material here for it to build up above the level of the water**. Once this has happened, the water returning to the sea as backwash at point 2 will deposit some of the sand as it runs out of energy (point C). This will also build up above sea-level in time. By this process **the beach extends itself into open water and is called a sandspit** (stage 2).

If a sandspit builds out into a bay, in time it might extend across the bay and join up with the beach on the other side (see Figure 19.4). When this happens, the coastal feature is **called a sandbar.** The shallow, stagnant seawater, trapped behind the sandbar, is called a lagoon. In time this will be filled in with wind-

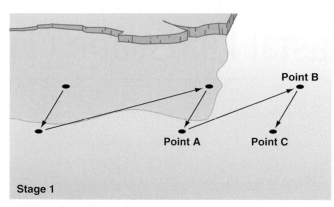

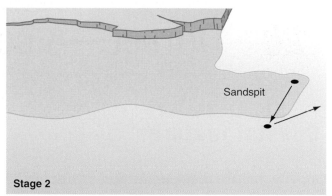

Figure 19.3 Formation of a sandspit

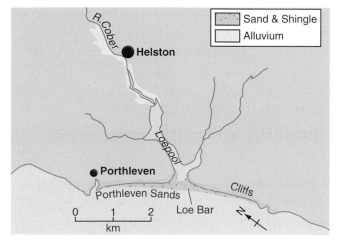

Figure 19.4 Loe Bar, Cornwall

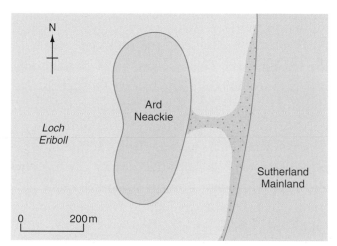

Figure 19.5 Tombolo in northern Scotland

blown deposits, be colonized by vegetation and eventually become dry land.

Along some coastlines, **a sandspit will grow outwards into open water and reach an island,** which it then joins to the mainland (see Figure 19.5). This is **called a tombolo.**

Although there are examples of sandbars and tombolos in the British Isles, **most sandspits do not reach very far into open water,** for two reasons. Firstly, the water is deeper further from the shore so it takes longer to build up material from the sea-bed. Secondly, many inlets have strong currents flowing into the sea (especially at river mouths), which will take away any material deposited by longshore drift before it can build up above sea-level.

QUESTIONS

1 Where are the largest and smallest deposits usually found on a beach? **(2)**

2 Explain how deposits on a beach become sorted **(4)**

3 Name three features formed by longshore drift **(3)**

4 a) Using a diagram(s), explain how a sandspit forms **(4)**
   b) Explain why sandspits form very slowly in open water **(2)**

5 What is the difference between a sandspit and a sandbar? **(1)**

6 Describe a tombolo **(2)**

Describe the landscape shown in Figure 19.2 (5)

20 Dorset—A coastal landscape (1)

Coastal features

Dorset is a small county on the south coast of England and contains some of the most striking examples of coastal features to be found anywhere in the British Isles.

As Figure 20.2 shows, several rock types reach the sea along the Dorset coast, some of them soft, others more resistant. The soft rocks, such as clay and shale, have been quickly eroded to form bays, such as Swanage Bay and Lulworth Cove. The resistant rocks, such as chalk and limestone, form cliffs and headlands that display caves, arches and stacks. Handfast Point is a good example of a headland and has stacks such as Old Harry and Old Harry's Wife.

The dominant winds and waves come from the south-west so the waves wash up the beach at an angle and longshore drift takes place. It has formed sandspits, such as Studland Spit, and the Isle of Portland has been joined to the mainland by a tombolo called Chesil Beach. It is 30 kilometres long and up to 15 metres high and is believed to have once been an offshore sandbar. Elsewhere, beaches are commonly found at the back of bays, such as Swanage Beach.

Tourist attractions

The Dorset coastline is very popular with tourists with 50 million visitor days every year. They are attracted partly by the natural landscape (in

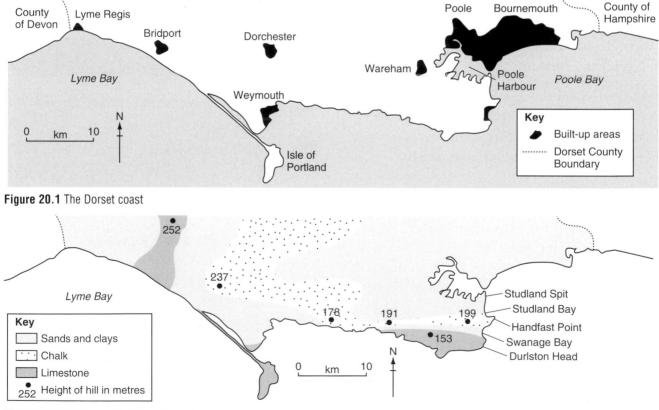

Figure 20.1 The Dorset coast

Figure 20.2 Geology of the Dorset coast

particular, the scenery and climate) and partly by the human landscape (its wide range of amenities and colourful history). Although the area has no motorway nearby, **it is still within two hours' drive of London and only one hour to Birmingham,** so it attracts many day-trippers as well as staying tourists.

The scenery

The coastline attracts many people who just come to sightsee. It has distinctive, white chalk cliffs and headlands that are studded with caves and arches and have strangely-shaped stacks in front of them. There are wide, sweeping bays, hidden, sheltered coves and long, sandy beaches.

The coast also attracts the more active visitor. The calm, sheltered water in the lagoons and bays and in Poole Harbour (the largest natural harbour in Britain) encourages **all sorts of watersport enthusiasts**—from swimmers to waterskiers to yachtsmen.

Figure 20.3 Coastal features

Just behind the coast, the sand dunes (e.g. at Studland), lagoons and marshes (e.g. The Fleet) are **important wildlife refuges** and this also encourages some visitors. Studland sand dunes contain rare heathland plants as well as rare British wildlife such as lizards and snakes. Poole Harbour is winter home for over 20 000 waterfowl.

The climate

Being on the south coast of England makes this one of the warmest areas in the country, with summer temperatures averaging 16° C (see Figure 20.4). The east-facing coasts here have the

distinction of being the sunniest places anywhere in Britain. Swanage has over 1700 hours of sunshine each year (compared with Glasgow's 1200 hours). The rainfall of about 800 mm per year is less than on the west coast but more than east coast resorts receive.

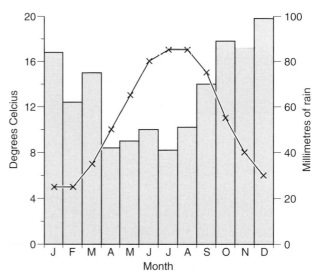

Figure 20.4 Climate graph of Swanage

QUESTIONS

1 Explain the effect of rock type on coastal features in Dorset (3)

2 Describe four coastal features that attract tourists to Dorset (2)

3 Explain why water activities are so popular in this area (2)

4 Where, along the Dorset coast, can wildlife be observed? (3)

5 Describe two advantages of the climate for tourism (2)

6 In what ways does Dorset's location explain its popularity? (2)

7* **There are many reasons why people go to Dorset for a holiday. List as many natural attractions of Dorset as you can find on these pages** (5)

21 Dorset—a coastal landscape (2)

The human attractions

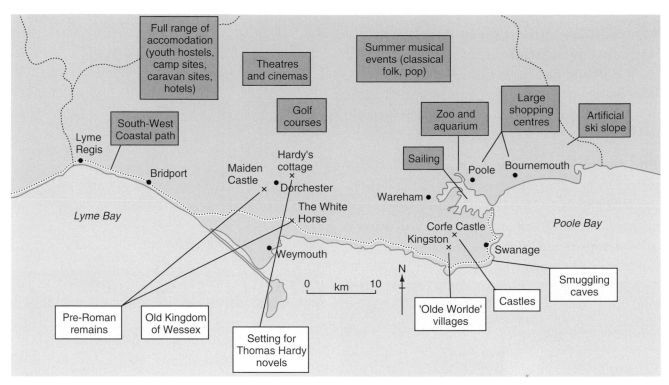

Figure 21.1 Tourist attractions in Dorset

QUESTIONS

1 Which of the human attractions of Dorset, Figure 21.1, would attract **a)** couples with young children, **b)** senior citizens? **(4)**

2 Which, do you think, attracts more visitors—the historical and cultural features or the amenities? Give reasons for your answer **(4)**

3 In your opinion, what are the three main reasons why people visit Dorset? Give reasons for your choices **(5)**

OS MAP QUESTIONS: On the OS Map of Swanage

4 Identify, with a grid reference **a)** caves, **b)** an arch, **c)** stacks **(3)**

5 Give a grid reference of an area with **a)** sea cliffs, **b)** a sandy beach, **c)** a shingle beach **(3)**

6 Using a diagram(s), explain how the headland of Peveril Point (square 0378) was formed **(4)**

7 Describe the natural attractions of the coast in the map area **(4)**

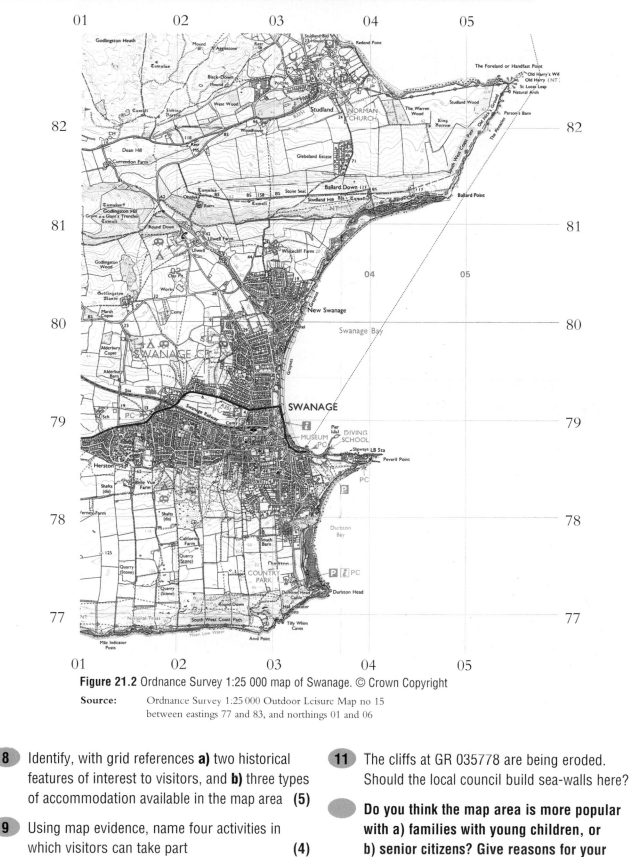

Figure 21.2 Ordnance Survey 1:25 000 map of Swanage. © Crown Copyright

Source: Ordnance Survey 1:25 000 Outdoor Leisure Map no 15
between eastings 77 and 83, and northings 01 and 06

8 Identify, with grid references **a)** two historical features of interest to visitors, and **b)** three types of accommodation available in the map area **(5)**

9 Using map evidence, name four activities in which visitors can take part **(4)**

10 What evidence is there that Swanage Bay suffers from longshore drift? **(1)**

11 The cliffs at GR 035778 are being eroded. Should the local council build sea-walls here?

Do you think the map area is more popular with a) families with young children, or b) senior citizens? Give reasons for your choice **(5)**

22 Dorset—a coastal landscape (3)

The impact and management of tourism

Increased employment and wealth

In 1997 **tourism provided 38,000 jobs in Dorset.** Because **more people have jobs they can afford to spend more money** and so local shops, restaurants and entertainments also benefit. The owners and managers of these services can then spend more money and so the extra wealth from tourism spreads throughout the community. This is called the **multiplier effect.**

Unfortunately **the number of tourists varies greatly from year to year.** In a poor year small villages are particularly affected because some of them rely on tourism for most of the jobs. Also, **many of the jobs are seasonal.** Staff are only needed for the summer months. So, although tourists bring jobs and wealth, they are not necessarily ideal jobs for all the local people. Larger centres, such as Bournemouth, **overcome the problem of seasonal jobs by attracting visitors all year.** They do this by providing conference facilities, encouraging student field-trips and promoting short-break winter holidays.

Traffic problems

Tourists cause a lot of congestion on the roads because (a) 82% of them travel to Dorset by car, (b) most come at the same time, e.g. Bank Holidays, weekends, and (c) they often drive slowly as they are sightseeing. **The larger settlements (Bournemouth, Poole, Swanage) all suffer from congestion and have tried various solutions,** e.g. **one-way systems,** encouraging other types of transport (by providing **bus lanes and cycle lanes**), **phasing traffic lights and restricting the hours of road works.**

The village of Corfe Castle also suffers badly. It is a major tourist attraction and half a million people visit it each year. It is also on the only main road to Swanage so a lot more tourists pass through on their way to the coast, as well as 500 heavy lorries every day. The congestion is also caused by the narrow streets, the street parking and the lack of proper car parks.

Dorset County Council have devised several solutions:

1. A railway line has been opened with steam trains running from Swanage to Norden, just north of Corfe Castle. People can now visit Corfe Castle and Swanage without driving into or through the village.
2. **An extra car park has been developed** in the village. This should reduce the street parking, which should allow traffic to go faster.
3. **More cycle ways** and **summer bus routes** have been developed, but a proposed by-pass has been rejected.

Honeypot problems

Poole Harbour is an example of a honeypot in Dorset, a place where large numbers of tourists visit. **There can be as many as 4000 boats in the harbour at any on time,** with people engaged in yachting, jetskiing, fishing, waterskiing and other water activities. Around the edge of Poole Harbour are sightseers, walkers, sunbathers and birdwatchers.

These tourists conflict with each other. The

noisy pursuits, e.g. powerboating upset the people who want peace and quiet, e.g. fishermen. Increasingly powerboats are using the harbour in winter, which is the time when thousands of birds migrate here.

To try and solve these conflicts Poole Council have introduced zoning of Poole Harbour. With this plan different activities are zoned in different areas so they do not upset each other (see Figure 22.1).

Maximum speed limits have been imposed in some parts, which prevents powerboating and waterskiing from taking place.

Landscape degradation

Degradation is when the landscape is spoiled or reduced in quality. **One type of degradation is the pollution of Poole Harbour.** This has been caused by oil from the many boats and from the small oilfield south of Poole Harbour, together with sewage from Poole itself.

Another example is the trampling of vegetation and erosion of sand dunes at Studland. Here, the dunes are between the car parks and the beach so, as people walk through the dunes, they trample the fragile vegetation until it dies. With less vegetation the sand dunes

themselves become eroded and wildlife disappears as the habitat changes. Horses also add to the trampling as there is a riding centre nearby.

QUESTIONS

1 Describe the changes in the numbers of tourists visiting Dorset during the year **(2)**

2 Suggest the types of jobs created by tourism **(4)**

3 Explain the multiplier effect of tourist jobs in this area **(2)**

4 What are the problems for a village in which most of the jobs are connected with tourism?

5 Tourists in Corfe Castle cause traffic congestion. What solutions has the village tried? **(3)**

6 Describe two examples of landscape degradation in Dorset **(5)**

7 Describe the problems that tourists bring to Poole Harbour **(4)**

8 What is meant by 'zoning' Poole Harbour and what benefits does it bring? **(3)**

9* **If you lived in a coastal village in Dorset, would you be pleased that it was a tourist resort? Give reasons for your answer** **(5)**

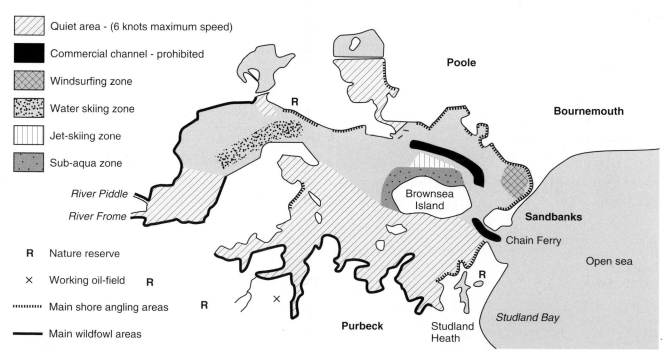

Quiet area - (6 knots maximum speed)

Commercial channel - prohibited

Windsurfing zone

Water skiing zone

Jet-skiing zone

Sub-aqua zone

River Piddle

River Frome

R Nature reserve

× Working oil-field

⋯⋯⋯ Main shore angling areas

——— Main wildfowl areas

Figure 22.1 The Management of Poole Harbour

23 Dorset—a coastal landscape (4)

Environmental conservation

Some of the tourist activity along the Dorset coast is not sustainable. If more and more people visit the honeypots in their cars and if more and more people use Poole Harbour, the landscape will soon become degraded. If this happens it will be less attractive for future tourists to enjoy. That is why **the environment here needs to be conserved**. Many organizations are involved in this and Figure 23.1 shows just some of the conservation areas that they have created.

Some of these areas are where **wildlife is protected** (e.g. RSPB sites), some where **vegetation is protected** (e.g. SSSI), some where

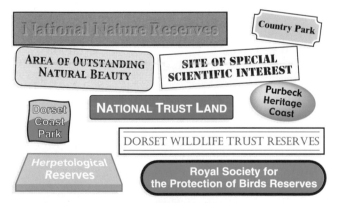

Figure 23.1 Conservation areas along the Dorset coast

Figure 23.2 A warden in Durlston Country Park talks to some local school children

Figure 23.3 Groynes in Bournemouth

buildings are conserved. (e.g. National Trust) and others where the **landscape as a whole is conserved** (e.g. Heritage Coast). They use a variety of methods. They **provide information and guided walks** to educate the visitors (e.g. Heritage Coast). They **restrict access** to very sensitive areas (e.g. National Nature Reserves). They **provide wardens** to look after the area (e.g. Country Park) and **they buy and manage land** themselves to ensure it is conserved (e.g. National Trust, Dorset Wildlife Trust land).

Environmental protection

The Dorset coastline is also being protected against natural forces. Sand is being moved along the coast by longshore drift. This upsets resorts that rely on their beaches to attract tourists. So, **town such as Swanage and Bournemouth have built wooden barriers down their beaches to stop longshore drift** (see Figure 23.3). These barriers are called **groynes**.

In many places the coastline needs protection from erosion. **Cliffs are being worn back** and this means that land is being lost, buildings destroyed and the cliff-tops are unsafe. At Osmington Mills, a small village that relies on tourism, cliff erosion threatens a caravan site, car park and cafe. To solve this problem, **concrete sea-walls are built**. They slow down erosion but they are

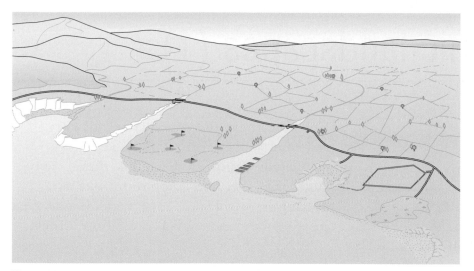

Figure 23.4 A coastal landscape

Figure 23.5

unattractive and reduce wildlife. Alternatively, **a beach can be built up in front of the cliff** to take the full force of the waves. One way of doing this is to construct a groyne nearby. The sand and shingle build up against the groyne and protect the cliff behind.

QUESTIONS

1 Explain why some tourism in Dorset is not sustainable (2)

2 Name four environmental conservation areas in Dorset and, for each, state what they aim to conserve (4)

3 Describe and explain three ways of conserving the environment in Dorset (6)

4 Describe the measures taken to reduce longshore drift in Swanage (3)

5 Describe the measures taken to reduce cliff erosion (2)

Techniques Questions

6 Figure 23.4 is a sketch of a coastal landscape. Draw the sketch and annotate (label) it to show the main landforms (3)

7 Draw a sketch of Figure 20.3 on p. 41 and label it to show the main coastal features (3)

Figure 23.5 shows a sketch of the centre of the village of Corfe Castle in Dorset. Draw this sketch and annotate it to show the main tourist attractions (5)

24 Volcanic landscapes (1)

Location

It has been at least 10 million years since any volcanoes were active in the British Isles, yet evidence of past volcanic activity can still be seen in many places. The main areas of volcanic rocks are shown in Figure 24.1.

Processes affecting volcanic landscapes

The shape of any landscape depends partly on the agents of erosion and the agents of weathering, but it also depends upon the nature of the rocks on the surface.

Volcanic rocks (e.g. basalt) are all very resistant, so the agents of erosion and weathering wear them down only slowly. This means that, over time, **they stand out as upland areas** while the softer rocks around them are worn away more quickly to form lowlands.

A volcanic plug

When a volcano becomes extinct, the molten rock or magma in its vent solidifies. This stops any more molten rock from reaching the surface (stage 1 in Figure 24.4). For this reason **the vent is now called a plug**, because it acts like an ordinary plug in filling a gap or hole.

Over time, the extinct volcano is subjected to erosion and weathering. As they are made partly of softer, looser ash, **the less resistant sides are eroded quickly. The plug is eroded slowly** as it is wholly made of magma. After millions of

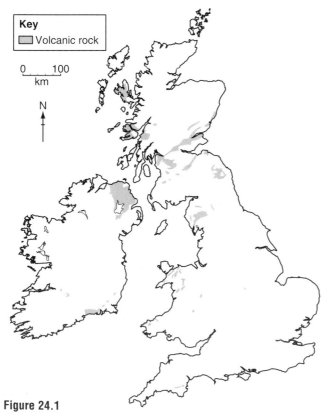

Key
▨ Volcanic rock

0 ___ 100
km

N

Figure 24.1

Figure 24.2 A volcanic plug

Figure 24.3

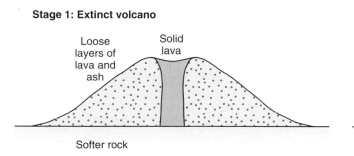

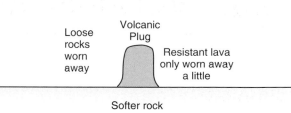

Figure 24.4 Stages in the formation of volcanic plug

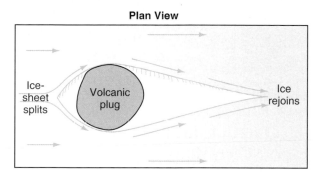

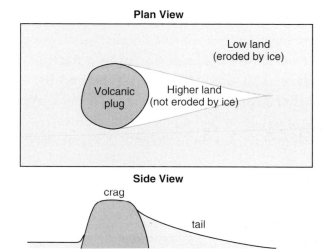

Figure 24.5 Stages in the formation of a crag and tail

years, the sides are worn away completely **and all that remains of the volcano is the volcanic plug**—a very steep-sided column of hard volcanic rock (stage 2).

A crag and tail

When ice-sheets spread over Britain during the Ice Age, they reached many volcanic plugs. Because a volcanic plug is very steep, **the ice-sheet split and went around the plug, joining together again behind it** (stage 1 in Figure 24.5).

Being very powerful, **the ice eroded the rock in front of the plug, but it could not erode the plug itself nor the land just behind the plug.** So, after the Ice Age, **the volcanic plug was left as a steep-sided hill, called a crag. Behind it**

is a sloping area of softer rock, not eroded by the ice, called its tail (stage 2). The whole feature is called a crag and tail.

QUESTIONS

1 Explain why volcanic rocks form upland areas **(2)**

2 Explain how a volcanic plug forms. You may use a diagram in your answer **(2)**

3 With the help of a diagram(s), explain how a volcanic plug becomes a crag and tail **(4)**

4* **Describe the features shown in Figures 24.2 and 24.3.** **(5)**

25 Volcanic landscapes (2)

A lava plateau

In some parts of the British Isles lava did not come to the surface at a single point (as a volcano) but, instead, the **lava erupted along a long crack or fissure** in the rocks. In these areas, huge amounts of lava flooded out across the land and slowly solidified, **forming a flattish surface**. This would have been **followed by another eruption**, in which another thick layer of lava would be laid down on top of the previous one, and so on (stage 1 in Figure 25.1)

Stage 1: Lava erupting at the surface

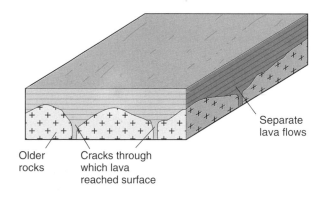

Stage 2: After thousands of years of erosion

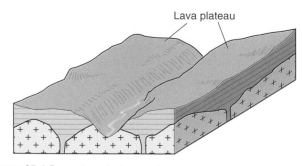

Figure 25.1 Formation of a lava plateau

Figure 25.2 Giant's Causeway in Northern Ireland

Being a hard rock, **the lava was eroded and weathered very slowly** so that, over the years, **it formed high land**. An area of flat, high land is called a plateau, so this feature is **called a lava plateau**—layers of lava forming flattish, high land.

In the millions of years since the lava plateau formed, rivers and ice have eroded them a little and frost and rain have weathered them a little, so the surface is not quite as flat and even as it was originally (stage 2 in Figure 25.1).

The largest lava plateau in the British Isles is the Antrim Plateau in Northern Ireland. It covers 4000 km². and is actually part of a larger plateau which extends to western Scotland. When the lava here cooled it formed six-sided columns, such as those found at the Giant's Causeway in Antrim (Figure 25.2) and the Isle Of Staffa in the Inner Hebrides.

Sills and dykes

Sometimes, during periods of volcanic activity, magma forced its way through cracks in the rocks but could not reach the surface. Where this happened there was no volcano or lava flow.

Stage 1: Magma enters rocks

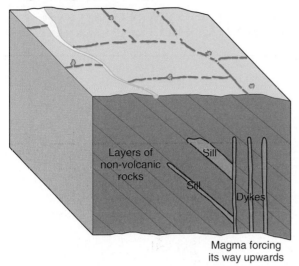

Layers of non-volcanic rocks

Sill

Sill

Dykes

Magma forcing its way upwards

Stage 2: After erosion

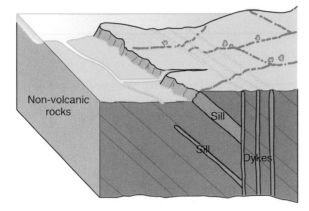

Non-volcanic rocks

Sill

Sill

Dykes

Figure 25.3 Formation of sills and dykes

Instead, the **magma spread out and cooled slowly below the surface**. Because it cooled slowly, it forms different types of rock, such as dolerite.

Where the magma spread out between the existing layers of rock and cooled down, **it formed a sill. Where the magma cut across the layers of rock** and cooled, **it is called a dyke** (see stage 1, Figure 25.3).

Over thousands of years the rocks above the sill and dyke are weathered and eroded away so that the sill and dyke can be seen at the surface. Because they are usually harder than the surrounding rocks, **sills and dykes are eroded**

more slowly and so form ridges (stage 2, Figure 25.3)

The longest sill in Britain is the Great Whin Sill in northern England. It forms a long ridge, along which part of Hadrian's Wall has been built. Where it cuts across the River Tees it causes High Force waterfall, the highest waterfall in England.

On the Isle of Arran, over 500 dykes can be seen along a 25-kilometre stretch of coast, forming a dyke swarm.

QUESTIONS

1. What is the difference between a volcano and a lava plateau **(1)**

2. Why does a lava plateau have layers? **(1)**

3. Why does lava form high, flattish land? **(2)**

4. Sills and dykes form underground. Explain how they become surface features **(2)**

5. Look at Figure 25.4. Which diagram (A or B) shows a dyke and which shows a sill? Give reasons for your answers **(2)**

6. Name one type of rock from which dykes and sills are made **(1)**

● **Describe what is taking place in stages 1 and 2 of Figure 25.1** **(5)**

Diagram A

Diagram B

 Volcanic rock

Figure 25.4

26 Edinburgh—a volcanic landscape (1)

Introduction

Edinburgh is located on the south side of the Firth of Forth within the Central Lowlands of Scotland. It is built on and surrounded by a host of volcanic features. The city's most famous landmark, its castle, sits atop a **volcanic plug**. This plug is also a **crag and tail** feature, the tail being the sloping land of the Royal Mile. The castle rock is just a side vent of a huge extinct volcano that erupted 350 million years ago. The main vent is now the volcanic plug of Arthur's Seat. Within a few kilometres of Edinburgh are other volcanic plugs, such as Bass Rock, North Berwick Law and Inchcape Rock.

Elsewhere, lava poured out of huge fissures in the ground building up **lava plateaus**. The Campsie Fells to the north-west and the Pentland Hills just to the south of Edinburgh formed in this way

Figure 26.2 Salisbury Crags

and have stepped sides caused by the different lava flows.

While this volcanic activity was taking place, not all of the molten rock reached the surface. Some spread out underground to form sills, such as Salisbury Crags and Corstorphine Hill.

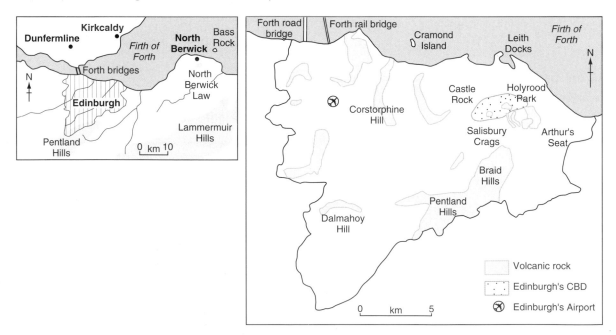

Figure 26.1 Location of Edinburgh

Salisbury Crags   Arthur's Seat   Castle Rock

Figure 26.3 Edinburgh's ancient volcano superimposed onto a view of the modern-day city

Tourist attractions

Natural attractions

Edinburgh is the only large city in the British Isles **built among hills**. It is also **beside the sea** and **has beaches** nearby, such as Portobello and Musselburgh. This gives it a unique setting that adds considerably to its appeal for visitors, even those who have come for the entertainments or on shopping trips. **The hills**, such as Arthur's Seat, **are popular walking areas** and, from their tops, **provide wonderful views** of the city below.

Unfortunately, Edinburgh is also one of the coldest cities in the British Isles, although this is counteracted in part by its **low rainfall** and **quite high sunshine total**. Nevertheless, Edinburgh's climate is unlikely to attract tourists.

Human attractions

Edinburgh's natural attractions are inseparable from its historical attractions. No matter which way you look in Edinburgh, you can see steeply-rising crags and hills and, on most of those hills, are ancient buildings of historic interest. It is this combination of human history and geological history that makes Edinburgh so attractive. The map on pages 54 and 55 shows the location of some of the many historic and cultural features of Edinburgh. It also shows some of the facilities or amenities that have been provided for tourists.

QUESTIONS

1. Name four volcanic features found in or near Edinburgh (4)

2. Describe the attractions of Edinburgh's scenery (3)

3. Describe the attractiveness of Edinburgh's climate for tourists (3)

4* Look at Figure 26.3. Describe the many ways in which the area around Edinburgh was different 350 million years ago (5)

27 Edinburgh—A volcanic landscape (2)

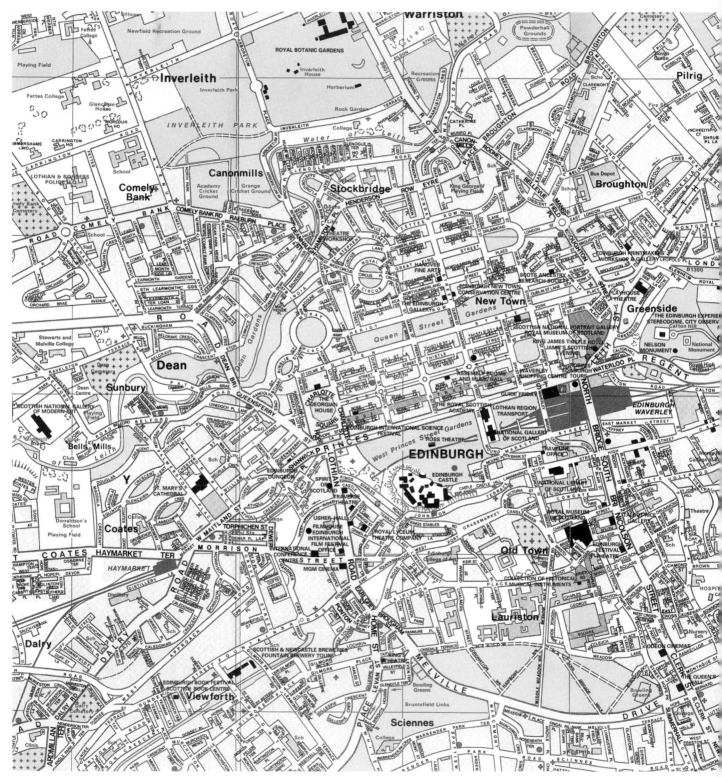

Figure 27.1

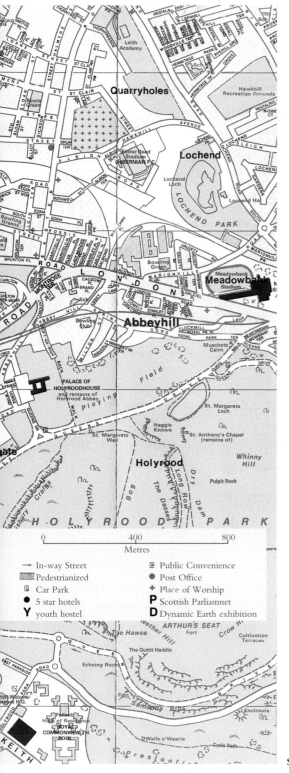

Source:　　Edinburgh Tourist Plan Pub. Estate Publications

QUESTIONS

The map shows many of Edinburgh's attractions

1 List six cultural attractions shown on the map　　　**(3)**

2 List four historical attractions　　　**(2)**

3 List six different amenities　　　**(3)**

4 Which **a)** cultural attraction, and **b)** historical attraction, do you think, attracts the most visitors? Give reasons for your answer　　　**(4)**

5 In what ways is the accommodation in Edinburgh suitable for different types of tourists?　　　**(2)**

The map shows many attractions in Edinburgh. List five attractions and explain why they are so popular　　　**(5)**

Figure 27.2 Various ameneties used by locals in Edinburgh are visible in this photograph

28 Edinburgh—a volcanic landscape (3)

The impact and management of tourism
Increased traffic

Two-thirds of Edinburgh's visitors come by car. Many of the others arrive by coach. About **50 000 coaches come every year** and **all the vehicles head for the 'honeypots'**, especially the area from The Royal Mile to Princes Street. This is also Edinburgh's Central Business District (CBD), so this area is already congested with the vehicles of local people going to work or shopping. The problem is made worse by **the lack of parking facilities**. It has been calculated that every motorist spends an average of 10 minutes looking for a parking place in Edinburgh. While doing so, he is driving slowly and holding up traffic even more.

The traffic **congestion** has become so bad that it **is putting off tourists from visiting the city** or making them leave early. This reduces the money they spend in the city. In addition, the congestion upsets the local people. It increases the number of accidents (there are nine serious accidents each year on The Royal Mile alone). It creates frustration among drivers and pedestrians and increases the air and noise pollution.

Edinburgh City Council have tried several solutions:

- **more coach parks** and signposted coach routes
- **park 'n' ride schemes** in the suburbs and **park 'n' walk** schemes at the edge of the CBD (e.g. at the St. James Centre)
- **linking pedestrian-only routes**, e.g. from the Royal Mile to Princes Street
- **traffic calming measures**, e.g. pavement widening, rumble strips, 'sleeping policemen'
- **building** new offices, leisure facilities and shopping centres in the suburbs, e.g. Gyle, Leith

Increased amenities

Because so many tourists visit Edinburgh, many amenities have been developed for them. Restaurants, shops and entertainments have all opened up especially to cater for tourists. But these amenities also benefit the local people. **Residents now have a much bigger choice of places to eat** (over 400 restaurants), **shop** (Harvey Nicholls, Jenners) **and relax** (nightclubs, casinos). They can also enjoy the many festivals, museums and exhibitions throughout the year. **Their quality of life has been improved** because of tourism.

Greater employment and wealth

The number of visitors to Edinburgh has steadily increased in recent years and, in 1997, reached 96 million visitor days. While in Edinburgh they spend £1500 million per year—on hotel bills, meals, souvenirs, taking trips etc. **This creates jobs**—as hotel receptionists, waiters, shop assistants, tour guides. The people with jobs then spend more money and so **the wealth is spread throughout the whole area**. This is called the **multiplier effect** and is shown in Figure 28.1.

In the past many tourist jobs were seasonal (for the summer only) but **now Edinburgh attracts people all year**. Hotels offer cheap short breaks in winter, e.g. shopping breaks, theatre breaks. They also provide facilities for companies to hold conferences and meetings in the off-peak season.

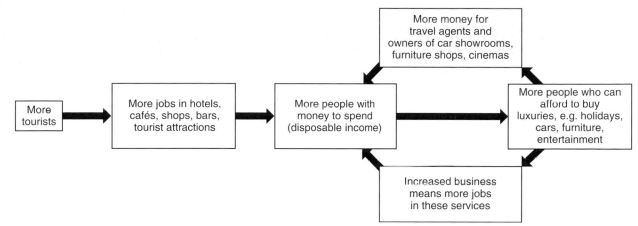

Figure 28.1 The multiplier effect of more tourists in Edinburgh

Edinburgh is now the 14th most popular conference destination in the world. Edinburgh City Council tries to arrange events for other times of the year, e.g. the Science Festival (April), Children's Festival (May) and, of course, the Hogmanay Festival. These help to keep staff employed throughout the year.

Pollution and landscape degradation

As the number of tourists have increased, so have the numbers of hotels, restaurants and entertainments. Figure 28.2 shows just some of the major developments in recent years. **Some of the new buildings** have won architectural awards (e.g. the Museum of Scotland). Others are less attractive and **spoil the views**. The tallest buildings also **obstruct the views** of the city. This **visual pollution may deter tourists** because many come especially to see the old buildings and their architecture. Other types of visual pollution in the city include litter, fly-posting and even hamburger stands.

Major tourist developments in

Edinburgh (1996–99)

Nine new hotels
Eight major extensions to hotels
Four conversions to hotels
Berthing of *Britannia*
Dynamic Earth Exhibition Centre
Museum of Scotland
Gallery of Modern Art
£100 million Leith waterfront development (including hotels, restaurants, shops, cinemas and a terminal for *HMS Britannia)*

Figure 28.2

QUESTIONS

1 Explain fully how tourists in Edinburgh add to the traffic congestion (4)

2 Describe the problems caused by traffic congestion (4)

3 What, do you think, is the purpose of traffic calming measures? (2)

4 Explain how new developments in Edinburgh's suburbs should reduce traffic congestion in the centre (2)

5 Describe the advantages of tourism to the local people (4)

6 What is meant by visual pollution and give three examples in the centre of Edinburgh (3)

7* **Edinburgh's main shopping street is Princes Street. Some people would like it to be pedestrianized (so no cars are allowed to drive there). What are the advantages and disadvantages of doing this?** (5)

29 Edinburgh—a volcanic landscape (4)

Environmental protection and conservation

Green belts

Edinburgh's green belt has been created partly to protect some of its most attractive environments. They include the volcanic plug of Arthur's Seat, the sill of Corstorphine Hill and part of the coastline near Cramond and Musselburgh (see Figure 29.1). The green belt was set up in 1957 and no development is allowed within it (except for farming, forestry and recreation) unless it is shown to be necessary. Unfortunately, the green belt area is also an attractive location for new housing, shopping centres, offices and the by-pass. If a company wishes to build its offices in the green belt because this is the best location for them and the company will bring many jobs and much money to the local area, it is very difficult for planners to reject its application. It is just as difficult to turn down applications for housing schemes and shopping centres if most people want to live and shop at the edge of the city. Over the years, therefore, planners have allowed some development inside the green belt. When they do this they add other areas to the green belt, so that it does not become smaller but just moves outwards.

Reducing air pollution

Many of the measures taken to reduce congestion will also reduce air pollution in Edinburgh's city centre. These include park 'n' ride schemes, traffic calming measures, the city by-pass and building shops and offices in the suburbs.

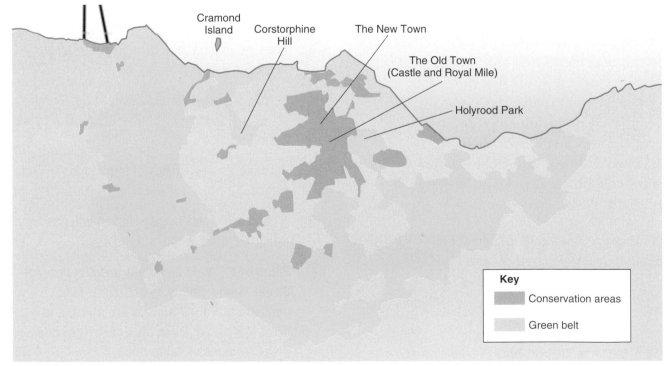

Cramond Island
Corstorphine Hill
The New Town
The Old Town (Castle and Royal Mile)
Holyrood Park

Key
Conservation areas
Green belt

Figure 29.1 Conservation and Green Belt areas in and around Edinburgh

Shop A: Acceptable shop front

Smaller windows   Small lettering   Elegant framework

Traditional building materials

Shop B: Unacceptable shop front

Large plate glass window   Over sized and clumsy lettering

Modern building materials

Figure 29.2 Acceptable and unacceptable shop fronts

Conservation areas and listed buildings

Conservation areas are areas of special architectural or historic interest. There are over 600 in Scotland and more than 20 in Edinburgh. Edinburgh, in fact, has more conserved buildings than any other settlement in Britain, except for London.

Within a Conservation Area the buildings are conserved by planning control. Anyone who wishes to make alterations to the outside of their building must ask for planning permission and it will be refused if it spoils the character or appearance of the area. This means that even the smallest changes can be refused. For example, in a Conservation Area it is not permitted to:

- fit cash dispensers to historic buildings
- have shop names in letters larger than 450 mm (see Figure 29.2)
- advertise on street furniture, even bus shelters in some areas
- have more than one flagpole!

It is necessary to ask for planning permission if you wish to:

- lop the trees in your garden
- paint the exterior of your house
- install a satellite dish

A Listed Building is one that is of particular architectural or historical interest. In the New Town of Edinburgh there are over 10 000 Listed Buildings. **Special planning permission has to be granted for any change to the outside and inside of a Listed Building**. The rules are even more strict and thorough. For example:

- your nameplate must be in bronze or brass and only cover one stone
- you might not be allowed to put alarm boxes on the fronts of buildings and if you are they must be painted the same colour as the stonework.

QUESTIONS

1 Name two areas in Edinburgh that are **a)** in the green belt, **b)** Conservation Areas   **(2)**

2 Explain how the environment is protected in a green belt   **(2)**

3 Explain why some parts of Edinburgh's green belt have been built on   **(4)**

4 What is protected in a Conservation Area?   **(2)**

5 Give examples of the rules that apply **a)** in a Conservation Area, **b)** to a Listed Building   **(3)**

Look at Figure 29.2. In parts of Edinburgh, the design of shop A is allowed but not that of shop B. Describe the differences between shop A and shop B   **(5)**

30   The Spanish Mediterranean Sea (1)

Landscapes under pressure

More people live in Europe than ever before. We move around more than ever before. We have more technology than ever before and we exploit our resources more than ever before. All these things put a strain on our land and water. And it is not just the urban areas where most of us live that suffer. Forests, mountains, wetlands, seas, coasts, rivers and lakes are all under increasing pressure. In some areas these landscapes are under such serious pressure that the quality of the environment there is becoming poorer. It is these areas that are studied in this topic and the first is our seas and coasts.

The Spanish Mediterranean Sea

'It took 3 billion years to make the Mediterranean Sea, but in 50 years we have almost destroyed it.'
(Jacques Cousteau, oceanographer)

Not all of the Mediterranean Sea is under environmental pressure. Not all of the western part around Spain is under pressure, but parts of it are showing signs of great strain (see Figure 30.1). This is not surprising as many of Spain's cities, factories and ports are here, there is a lot of intensive farming and millions of people visit here each year on holiday. The impact of industry and tourism have been particularly severe and these are now studied in more detail.

Figure 30.1 Seas and coasts under serious pressure (shaded in grey)

Pressures related to industry

Although it is more well-known as a holiday region, the Mediterranean coast of Spain has several industrial areas (see Figure 31.1). These pollute the environment in various ways.

Water pollution from heavy metals

Tiny particles of waste metal (e.g. mercury, lead, iron, aluminium) from factories and mines along the Mediterranean coast are dumped or leak into the sea (see Figure 30.2). They **are absorbed by fish, which are then eaten by humans**. If large amounts of these metals build up in people **they can cause illness, including cancers**.

The **pollution is also killing many marine plants**, which is allowing alien seaweed to invade the waters around the Balearic Islands. The seaweed contains a toxin that kills algae.

Oil pollution

Oil is dumped into the sea by chemical works, oil refineries, tankers and other ships. The coast

around Barcelona is particularly badly affected. **The oil kills fish** because it is toxic and also because it coats their gills. When the oil sinks to the sea-bed, **it then harms fish sperm** so that there will be fewer fish in the future.

Solutions to industrial pressures

1. **European Union** countries have **banned the dumping of some metals**, e.g. mercury and cadmium, and have **strict controls over others**. This was part of the 1975 Mediterranean Action Plan.

2. **Environmental pressure groups have been involved in direct action**. For example, in 1998, Greenpeace ecowarriors scaled the high chimney of a Barcelona factory to protest at its dumping of toxic industrial waste. They also sent their own ship to help clean up after the 1998 leak from a zinc mine (see Figure 30.2).

QUESTIONS

1. Name four seas in Europe under environmental pressure **(2)**

2. Look at Figure 31.1. Describe the location of the industrial areas along the Mediterranean coast of Spain **(2)**

3. Name two types of industrial pollution **(1)**

4. Describe two ways in which industrial pollution harms the Mediterranean environment **(4)**

5. In what ways do (a) the European Union and (b) environmental groups try to control industrial pollution? **(4)**

6*. **Do you think it is possible to stop factories from polluting the sea? What is the best way of stopping them?** **(5)**

26 April, 1998

Deadly Toxic Spill in Southern Spain

It has been reported that 5 million cubic metres of poisonous liquid waste escaped from a zinc mine in south-west Spain yesterday. It has already covered 6000 ha of farmland, killing crops and poisoning the soil for years to come. The water and sludge, containing arsenic, lead and cadmium, is now spreading along the coast towards the Coto Donana National Park, leaving a trail of dead fish. As the poisonous heavy metals get into the food chain, all the local wildlife will be at risk, including the rare lynx and royal eagle found here.

Greenpeace has sent its environmental flagship to assess the damage.

Figure 30.2 News stories like this are becoming more common

31 The Spanish Mediterranean Sea (2)

Pressures related to tourism

The Mediterranean coast of Spain used to be an area of farming and fishing with a few larger settlements, such as Barcelona and Valencia. This changed in the 1960s with the arrival of mass tourism. Now, 60 million people visit Spain each year and the vast majority head for the Mediterranean. Small fishing villages have mushroomed into noisy resorts that stretch for several kilometres along the coast, swallowing up farmland for roads, hotels, shops, discos and restaurants.

Such a rapid growth in tourism has put a strain on the local environment. High-rise apartment blocks now hide the view of the hills behind the coast. Traffic has increased enormously, as have crime-rates, litter and pollution. Two of these problems are now examined in more detail.

Water pollution from sewage

A huge rise in the number of tourists staying along the coast means a great increase in the amount of sewage dumped into the Mediterranean. Some towns still dump sewage raw into the sea. This causes several problems.

Raw sewage spreads disease, either (a) by swimmers coming into contact with it, or (b) by people eating shellfish that have been contaminated with the sewage. Diseases such as skin fungus, hepatitis, typhoid and gastro-enteritis are spread in this way.

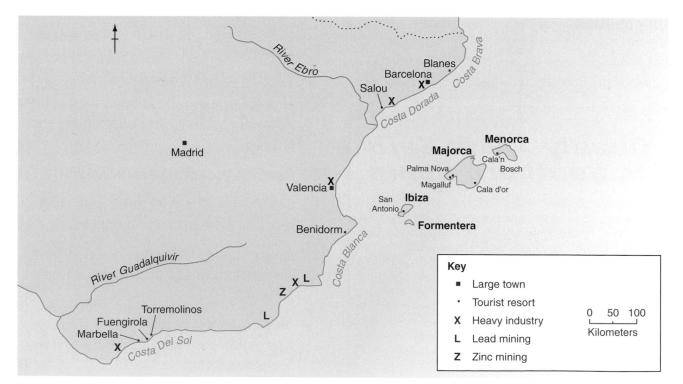

Figure 31.1 The Spanish Mediterranean

Raw sewage is also an 'oxygen-robber'. It needs a lot of oxygen to decompose, which means less oxygen for fish and other marine life. As a result, the amount of wildlife in the sea is reduced.

Water extraction

More tourists need more water, not just for drinking but for growing the fruit and vegetables they eat and for watering the golf courses they use. But Mediterranean Spain receives little rain, especially in summer. To overcome this, water is taken from the few rivers in the region. This, however, also causes problems.

Putting dams across the River Ebro has stopped silt from going downstream. This means that its delta, made from the river's silt, is shrinking and it is an important wildlife habitat.

Also, with less water flowing into the sea, the sea can now reach further inland at high tide. But the freshwater plants and animals cannot survive in saltwater and are dying out or moving away.

Solutions to tourist pressures

1. The **European Union** awards *blue flags* to beaches that have a high standard of water

Majorca Running Out of Water

Majorca has no rivers. It has two reservoirs which are rarely half full. It also has underground water, but the level of this is falling by up to 5 metres a year. So, with little rain and more and more tourists, Majorca is desperate for water. It has begun to bring water from the River Ebro to the island by tanker, but there are objections to this

Faced with this, and other problems caused by tourism, the Majorca authorities are thinking of taxing tourists. This would increase holiday prices by 5–7%. They would use the money to fund environmental projects on the island.

Figure 31.2

quality and beach cleanliness. Of the 597 beaches inspected in Spain, 364 were awarded blue flags in 2000.

It **issues guidelines**, advising countries that all sewage should be treated and that it should be dumped at sea beyond the swimming limits and the shellfish breeding waters.

2. **Environmental groups have well-publicized campaigns**, e.g. for *sustainable tourism*. This type of tourism ensures that the local environment is not harmed, so that future generations can also enjoy its attractions. By polluting the sea, reducing the wildlife and causing visual and noise pollution, tourists are harming the environment so that future generations will not wish to come. Groups such as Greenpeace highlight the damage caused by tourism. **They also name and shame** those beaches that are still very polluted.

QUESTIONS

1. Name two tourist areas in Mediterranean Spain **(1)**

2. Describe ways in which tourists can pollute the Mediterranean Sea **(6)**

3. Explain how tourists in Spain can cause water shortages **(4)**

4. Explain how damming the River Ebro reduces wildlife downstream **(4)**

5. Do you think the European Union's methods of reducing environmental pressures caused by tourism are likely to be very effective? Give reasons for your answer **(4)**

6. a) What is 'sustainable tourism'? **(2)**
   b) In what ways is tourism in the Mediterranean *not* sustainable? **(2)**

   Look at Figure 31.2. Do you think Majorca should tax tourists? Give reasons for your answer **(5)**

32 The River Rhine (1)

Rivers and lakes under serious pressure

Rivers have always been important to people—as a source of food, for drinking water, a means of transport and even a means of defence. But as well as using rivers and lakes, we also abuse them. We use them to dump sewage, industrial waste and farm chemicals and they also become polluted with oil from boats. As a result, parts of some of our rivers are now under serious environmental pressure. They are shown in Figure 32.1 and the state of one of them—the River Rhine—is now examined in more detail.

The River Rhine

Figure 33.1 on p. 66 shows the 1320-kilometre course of the Rhine from its source in the Swiss Alps to its mouth in the Netherlands. In between, it passes through Germany and has borders with France, Austria and Liechtenstein. Fifty million people live within its drainage basin. Europe's largest coalfield and industrial area is on its banks. It is also Europe's busiest river, with 10 000 vessels using it at any one time, and one of the world's busiest ports is at its mouth. It is hardly surprising, therefore, that the river is under serious environmental pressure, and this pressure comes chiefly from industry and urban areas.

Pressures related to urban areas

The many towns along the banks of the Rhine have interfered with it in many ways over many years. They have straightened it, deepened it, dammed it, polluted it and taken water from it. All of these activities affect the river environment and the wildlife to be found there.

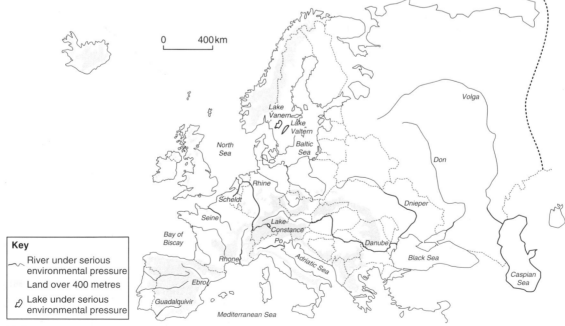

Key
— River under serious environmental pressure
  Land over 400 metres
⌂ Lake under serious environmental pressure

Figure 32.1

Water pollution from sewage

Sewage affects the Rhine in the same ways as it affects the Mediterranean Sea (page 62). **Untreated sewage is foul-smelling and carries bacteria**, which spread infection. **Treated sewage is an oxygen-robber**, which results in some wildlife in the river either dying out or moving away. Salmon, for example, were very common in the Rhine until the 1950s, but have not been seen since.

Water pollution from landfill sites

The 50 million people living near the Rhine create a lot of rubbish, most of which is dumped in landfill sites (Figure 32.2). These are holes in the ground in which waste is spread in layers and covered with soil. Many landfill sites contain unsorted rubbish, which may include toxic chemicals. Unless lined properly, **rainwater can drain through the landfill site and take chemicals away in solution**. This water then drains into the Rhine, adding to its pollution.

Figure 32.2 A landfill site in Germany

Solutions to pressures from urban areas

1. The **European Union has issued guidelines** to its member countries **on sewage disposal**. In addition, the European Union countries bordering the Rhine formed an association in the 1950s, called the International Commission for the Protection of The Rhine (ICPR). **They made laws** forcing local authorities to improve their sewers and build more sewage-treatment plants. Their target for the year 2000 was that 90% of all sewage entering the Rhine would be treated.

2. **Environmental pressure groups, such as Friends of the Earth, have campaigns against landfill sites**. They campaign for more recycling of rubbish (Figure 32.3) and put forward alternative, environmentally-friendly ways of disposing of organic waste, e.g. composting and anaerobic digestion. Partly because of these campaigns, the number of alternative disposal sites in Germany and the Netherlands is growing.

Figure 32.3 Recycling can be a better way of dealing with rubbish

QUESTIONS

1   Name four rivers and two lakes in mainland Europe under serious environmental pressure **(2)**

2   a) How many people live within the drainage basin of the River Rhine?
    b) Using Figure 33.1, name four cities on the banks of the Rhine **(2)**

3   a) What is a 'landfill site'?
    b) How do landfill sites pollute the Rhine? **(4)**

4   Describe other ways in which urban areas pollute the Rhine **(4)**

5   Describe ways in which European Union countries reduce pollution in the Rhine from urban areas **(4)**

6* **Friends of the Earth say that everyone should produce less rubbish. This would then reduce pollution in rivers. Suggest ways in which people can do this.** **(5)**

33   The River Rhine (2)

Pressures from industry

About 70% of all the industries in the Netherlands, Germany and Switzerland are located near the river Rhine (Figure 33.1) and some use the river to dump their waste. This has resulted in a grossly polluted river in places which, in the 1970s, hardly supported any forms of water life.

Water pollution from industrial chemicals

Industries, such as paper making, brewing, chemicals and detergents, regularly **dump their waste into the Rhine**. This waste includes pollutants such as cadmium, mercury, lead, phosphates, nitrates and salt. Many of the more **sensitive species of fish and other water life cannot survive** in these polluted waters and have long since died out. **Fewer species now remain.**

In addition, frequent **accidental spillages occur**. For instance, in 1986, a fire in a chemical works in Basle, Switzerland, led to 30 tonnes of assorted chemicals pouring into the Rhine. They formed a thick red slick in the river that flowed slowly downstream, 1200 kilometres to the sea. The slick was so poisonous that it killed half a million fish and river birds and even sheep that drank from the river. It killed the tiny water fleas at the base of the food chain, so reducing future water life. On finally entering the North Sea many days later, the chemicals ruined the important oyster beds there.

Thermal pollution

Some factories and most **power stations dump liquid waste in the Rhine, which is many**

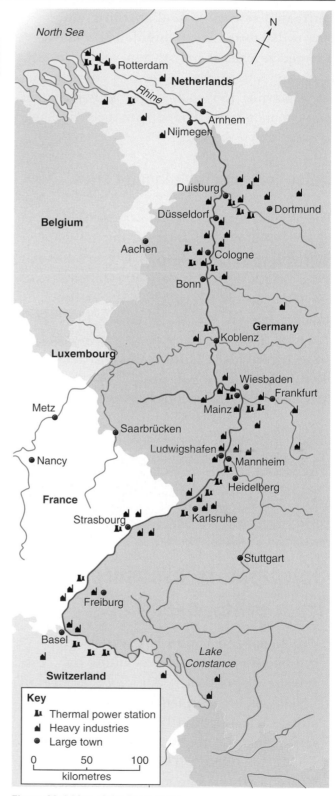

Figure 33.1 Map of the River Rhine

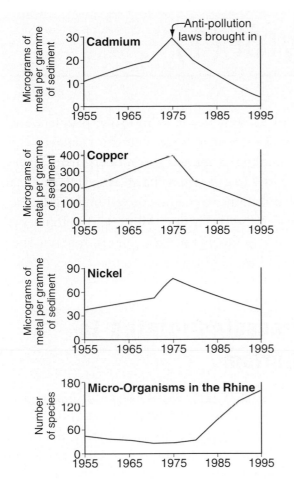

Figure 33.2 Changes in the quality of the Rhine (1955–1995)

degrees warmer than the river water (called *thermal pollution*). This sudden rise in temperature **can be fatal to the larvae of some river wildlife and to the tiniest river fleas** at the base of the food chain. As their numbers decrease, there is less food for other wildlife and their numbers decrease also. Some species of fish e.g. trout, move away whereas more tolerant species, such as pike, move in.

Warm water also confuses fish and causes their eggs to hatch out in winter when there is no food for them to survive.

Solutions to industrial pressures

1. The **European Union** countries bordering the Rhine formed the ICPR which **brought in** **strict laws to reduce pollution**, forcing industries to treat their waste before dumping it. Figure 33.2 shows how successful these laws have been, with the number of pollutants decreasing and the number of species increasing. The European Union **also allows companies** whose goods are produced by environmentally-friendly processes **to have ecolabels on their products**.

2. **Environmental groups may try to solve the problems by direct action**. For example, in September 1999, Greenpeace ecowarriors boarded a barge at the mouth of the Rhine and successfully stopped it from dumping toxic waste. As pictures of this were shown on TV and in newspapers all over Europe, they managed to bring the problem to the attention of millions of people.

QUESTIONS

1   **a)** Name four types of industry that pollute the river Rhine

     b) Name four chemicals commonly dumped in the Rhine   **(4)**

2   Describe the effects of chemicals on the river environment   **(5)**

3   Chemicals and warm water kill water fleas at the base of the food chain. Explain how this affects other wildlife in the river   **(3)**

4   Describe two other environmental effects of thermal pollution   **(2)**

5   Describe the changes shown in Figure 33.2   **(3)**

6   Explain how European Union countries have tried to reduce industrial pollution in the Rhine   **(4)**

Environmental groups are involved in direct action. Do you think this will greatly reduce pollution in the Rhine? Give reasons for your answer   **(5)**

34 The Austrian Tyrol (1)

Mountains under serious pressure

For hundreds of years, Europe's mountains were remote areas in which few people lived and which very few ever visited. This began to change 50 years ago, as our technology improved and roads and railways were built over and under the mountains. More people came to live and to visit, but the mountain environment is a fragile one and increasing numbers of residents, tourists and their vehicles have put some of these regions under serious pressure. The Alps, being the highest and most accessible mountains, are beginning to suffer badly.

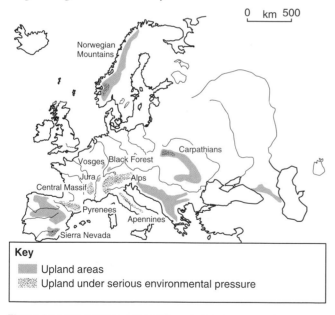

Figure 34.1 Mountains under serious pressure

The Tyrol region of Austria

The Alps cover 70% of Austria and the highest parts are in the province of Tyrol. Here there are small snowfields and glaciers and the mountains

rise to 3774 metres. Traditionally, people were dairy farmers or foresters, but this changed with the building of tunnels and modern roads over the passes (e.g. Brenner Pass). Also, hydro-electric power stations were constructed high up in the valley of the River Inn. All these developments opened up the Tyrol, especially to tourists, and have led to increasing environmental pressures here.

Pressures related to tourism

Until 40 years ago, tourism in the Alps was for the very wealthy or the very adventurous. With the arrival of package holidays, the number of tourists has increased dramatically and the Tyrol is finding it difficult to cope.

Visual pollution

So much development has taken place in the most popular parts of the Tyrol **that the scenery is being spoiled** (see Figure 34.2). In almost every village there are **new hotels, guest houses and restaurants** and, because there are now more jobs and more people, **more houses** have also had to be built. Away from the villages, the mountainsides are now covered with skiing facilities, such as **chairlifts and cable cars**. The character and landscape of the area has changed rapidly and local people are concerned that, in future, fewer tourists will wish to visit the area, affecting the jobs of thousands of people. It is worse in the off-peak season, when many of the shops are shut, the buildings are empty, the chairlifts are still, the people who work there have gone away and there is a 'dead' atmosphere.

Figure 34.2 An unplanned landscape in the Tyrol

Deforestation

The slopes of the Alps here are forested with pine, spruce and larch, but **these trees are being cut down at a rapid rate to make way for tourist developments**, ski runs, better roads and to supply building materials and fuel to the growing population.

Tree roots bind the soil and rocks together and prevent snow and ice from slipping. **With fewer trees, there are more landslides and avalanches**. In February 1999, for instance, a 300 km/h avalanche killed 31 people in the resort of Galtur.

Trees also intercept rain and absorb water so that less runs off into rivers. **With fewer trees, there is more run-off and rivers flood** more often. Floods in recent years have killed many people and wrecked homes and holiday areas.

Solutions to tourist pressures

Before any major project takes place **in the European Union, an environmental impact assessment must be completed**. The likely effects of the project on the flora, fauna, soil, water, air and landscape are assessed, as well as ways of reducing the environmental impact. At present, this is done just to inform the developers. It is not possible to stop projects that are environmentally damaging.

Environmental groups campaign for sustainable tourism (see page 63). They campaign for greater planning in tourist areas, so the scenery is not spoiled and villages retain their character. They also campaign for local people to be involved in the planning, as they are the people most interested in sustaining tourism for their children and grandchildren.

QUESTIONS

1 Name four mountain regions in Europe under serious environmental pressure **(2)**

2 Look at Figure 35.1.

   a) Name the largest town in the Tyrol
   b) How high are the Alps here?
   c) Name one river flowing through the Tyrol **(2)**

3 **a)** What is 'visual pollution'?
   b) In what ways are tourists responsible for visual pollution in the Tyrol? **(4)**

4 Explain how cutting down trees can cause
   a) avalanches, and **b)** flooding **(6)**

5 Explain how environmental groups try to reduce tourist pressure in the Alps **(3)**

6 **a)** What is an 'environmental impact assessment'?
   b) How effective is it? **(5)**

7* Do you think the people of the Tyrol are pleased that so many tourists visit there? Give reasons for your answer **(5)**

35 The Austrian Tyrol (2)

Pressures on agriculture

Pollution from traffic

The Tyrol (Figure 35.1) lies between the industrial regions of south Germany and north Italy, so traffic is busy along the main routes such as the Brenner motorway. Road traffic has increased five-fold over the last 30 years and the lorries have also increased in size. The **noise pollution caused by the heavy traffic** is made worse because it echoes around the mountain valleys. There is **air pollution from vehicle exhausts**, which release carbon monoxide, sulphur dioxide and nitrogen oxides into the air. This air pollution cannot escape easily from the steep valleys, especially in winter when there is high pressure and calm conditions. **The air is also polluted with salt** that has been spread on roads. Winds pick up the salt particles and may deposit them on trees, which can kill them.

Changes to the farm landscape

Figure 35.2 shows some of the changes that are taking place in the Tyrol. Until recently, most of the people here were farmers who lived in villages in the valley floor but stayed with their cattle, sheep and goats in huts on the middle and high pastures during the summer. Here they made hay and turned milk into cheese. By making use of the slopes in summer, the best land in the valley floor could be used for growing fodder crops for the winter. This is called *transhumance farming*.

But now, with more alternative jobs in the countryside and, especially in the towns, **there are fewer people who wish to work the land** and there are not enough farmworkers to carry on transhumance farming. Instead the **farmers' huts are taken over as chalets by tourists**. Small **tourist villages have even grown up** on these summer pastures.

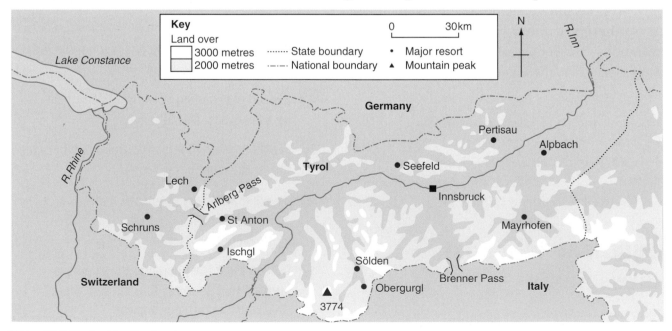

Figure 35.1 The Austrian Tyrol region

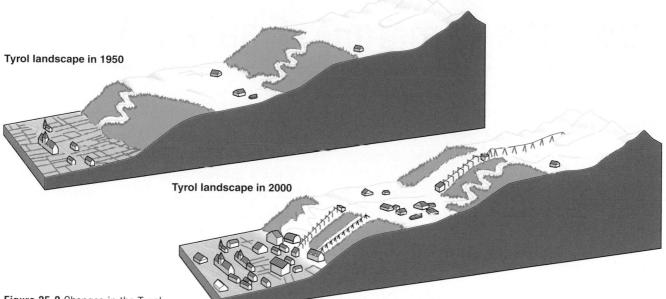

Tyrol landscape in 1950

Tyrol landscape in 2000

Figure 35.2 Changes in the Tyrol

The area of **farmland in the valleys is also decreasing** because of new tourist developments, housing schemes, recreational facilities and new roads. Dairy farming is still important, but more land is now used for growing fruit and vegetables for the local population. The traditional way of life and the landscape is changing.

Solutions to agricultural pressures

The **European Union countries have made laws** limiting the amount of noise that lorries can make. Lorries with a green disc and a letter **L** are those that conform with the EU noise regulations. In addition, lorries are not allowed to drive through Austria at night or on Sundays or on Saturday afternoons. The EU have also brought in laws limiting the amount of air pollution from vehicles. To try and maintain the farming way of life here, **the EU give grants for farm improvements and subsidies** for every animal owned. By making farming more profitable, they hope more people will stay and farm mountain regions such as the Tyrol.

Environmental groups have tried direct action. For example, in October 1996, Greenpeace activists blocked the Tauern Highway in the Tyrol to demonstrate against air pollution caused by lorries. They want the government to bring in higher tolls so that fewer vehicles use the roads.

QUESTIONS

1 In what ways do local conditions make air and noise pollution worse in the Tyrol? **(3)**

2 Describe four changes in the farming landscape since 1950, shown in Figure 35.2 **(4)**

3 Explain why fewer farmers in the Tyrol now farm the higher pastures **(3)**

4 Explain why the area of farmland is decreasing so rapidly here **(3)**

5 Describe two ways in which the European Union try to reduce pollution from traffic **(2)**

Which is a better way of reducing traffic noise: charging tolls or reducing lorry size? Give reasons for your answer **(5)**

36   The Camargue, France (1)

Wetlands under serious pressure

Wetlands are areas where water sometimes covers the land. They are found at the coast and in river floodplains. Because they are part water and part land, they attract a great variety of wildlife. But, if there are slight changes to this delicate balance between land and water, environmental pressures start to build up. Figure 36.1 shows the wetland areas where the environmental pressures have become quite serious. The Camargue, in south France, is one of those areas.

The Camargue wetlands

The Camargue is the delta of the River Rhone, where it enters the Mediterranean Sea. It is a triangular-shaped area of 850 km^2 between the Grand Rhone and the Petit Rhone (see Figure 37.1). It is a mixture of saltwater and freshwater, marshes, lagoons and dry land. It is watery in winter but dry and salty in summer. The area is famous for its special wildlife—its pink flamingoes, wild horses, black bulls and thousands of migratory birds. But it is also a living, working environment and this gives rise to environmental pressures, especially from tourism, industry and agriculture.

Pressures relating to agriculture

Nearly half of the Camargue is farmland and over half of the people who live here earn their living from farming. Farmland is also expanding and

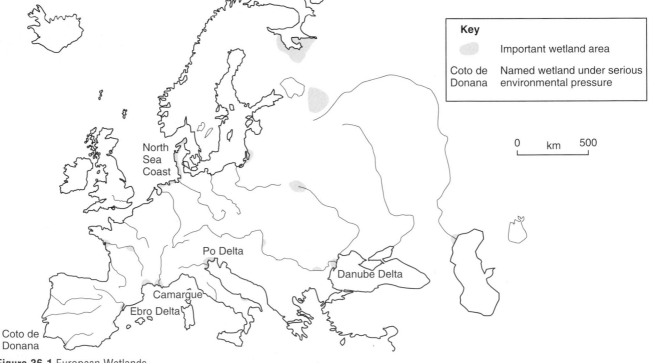

Figure 36.1 European Wetlands

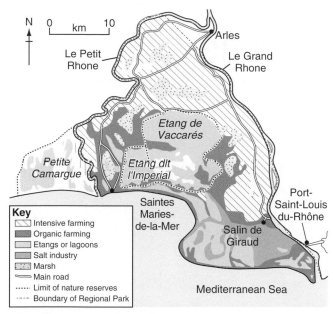

Figure 36.2 Zones within the Camargue

taking over wetland areas. Half of the Camargue's natural wetlands have disappeared in the last 50 years. Farmers grow wheat, vines, sunflowers and rice under irrigation, but intensive farming brings problems.

Pollution from agricultural chemicals

Many of the farms in the Camargue use fertilizers and pesticides to increase their crop yields. These chemicals are dissolved by rain and either seep into the soil and rock or run–off over the surface into the nearest river or lagoon. If the **pesticides build up in water,** they are toxic **and will kill the tiny water life** at the base of the food chain. **Fertilisers,** on the other hand, **help tiny plants, called algae, to grow.** As they multiply, they use up more oxygen in the water, **causing fish and other water life to suffocate.** Polluted water is one reason why some of the Camargue's salt pans have been abandoned.

Drainage schemes

The Camargue soils are generally quite fertile because the waters of the Rhone regularly flood the area and leave behind a covering of rich sediment. Once drained, therefore, these soils can be used for growing crops. But, **draining the**

water from the land means that water plants (such as reeds) die out, as do the wildlife that depend upon them. Also, with less water, the land now floods less often. So **the soil does not receive deposits of rich sediment** and becomes less fertile.

Solutions to agricultural pressures

The **Camargue Regional Park was set up** in 1972 to manage the environmental pressures here. It is run by a Foundation, which is **made up of people** from all sections of the community, **including environmental groups, and it also receives help from the European Union.** To avoid conflicts between land users, **the different activities in the Camargue are zoned** (see Figure 36.2). Separate areas are set aside for industry, tourism, agriculture and nature reserves. The two nature reserves are especially sensitive areas, including the flamingo breeding grounds, in which no-one lives and no farming is allowed. Outside the reserves is a zone in which only organic farming can take place.

QUESTIONS

1. Name four wetlands in Europe   (2)

2. Describe the natural landscape of the Camargue wetland   (4)

3. Explain how **a)** fertilizers, and **b)** pesticides pollute the Camargue   (6)

4. Explain why Camargue soils are becoming poorer   (3)

5. Who runs the Camargue Regional Park?   (2)

6. Explain how zoning reduces the problems caused by farming activities   (3)

7.* **Do you think that people should be allowed to live and work in the Camargue? Give reasons for your answer**   (5)

37 The Camargue, France (2)

Pressures relating to industry

Land and visual pollution

The Camargue is the most important producer of sea salt in Europe. The industry covers a large area in the south-east of the region. Seawater is pumped into lagoons and left to evaporate, leaving behind salt deposits. But the salt seeps into soils nearby, which reduces plant life and lowers crop yields.

The Camargue is surrounded by industries and towns that get closer every year. As well as the salt works in the south, there is the expanding town of Arles to the north, tourist developments at La Grande Motte to the west and the biggest port in France (Marseilles—Fos) just to the east.

These are very visible and cause visual pollution, especially the white salt works (see Figure 37.2).

Figure 37.2 A salt works

Air pollution

Air pollution in the Camargue comes from the many industries and power stations that surround it and from the heavy lorries that take away the sea salt. **Factories, power stations and motor vehicles** burn fossil fuels (coal, oil, gas), which releases carbon dioxide into the atmosphere. Carbon dioxide is a greenhouse gas,

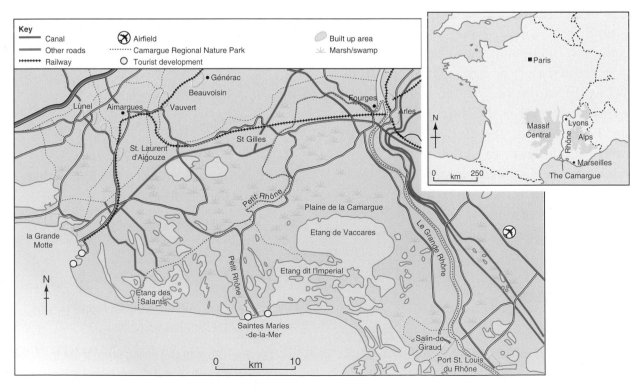

Figure 37.1 The Camargue region

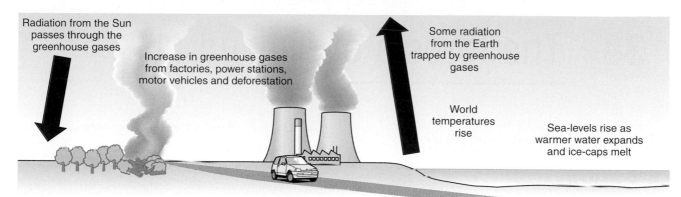

Figure 37.3 The greenhouse effect

which means that it allows through energy from the Sun but absorbs heat from the Earth. (This is the greenhouse effect). So, as carbon dioxide builds up in the atmosphere (30% more than 200 years ago), the temperature of the air and the Earth rises. This is global warming, and has led to an increase in temperatures of 0.5°C in the last 100 years.

As global warming takes place, sea-levels rise (a) because some of the ice at the Poles melts, and (b) because water expands when it heats up. Experts estimate that global temperatures may rise by 2°C by the year 2100 and that **the Mediterranean Sea will rise by 1 metre.**

A rise in sea-level would be devastating to the Camargue. Some of it would be drowned completely, whereas the freshwater areas would be invaded by salt water. Many of the plants and animals would disappear, as would the salt works and some of the farmland.

Solutions to industrial pressures

European Union countries are responsible for one-sixth of all the carbon dioxide that is emitted into the air. After the Kyoto Climate Conference of 1997, **EU countries agreed to cut carbon dioxide emissions** to 5% *less* than 1990 levels by the year 2010. Unfortunately, a reduction in carbon dioxide will take 50 years to affect sea-levels. **The EU has over 200 environmental laws, which put limits on air pollution** from factories and motor vehicles.

Many **environmental groups have had campaigns**, warning people of the dangers caused by global warming, long before the European Union took action. Also, **environmental groups within the Camargue Regional Park Foundation help to reduce land and visual pollution by zoning industry** away from sensitive habitats and from farmland. **They try to protect the scenery,** even insisting that all electricity cables and telephone wires be placed underground.

QUESTIONS

1 In what ways does the salt industry pollute the Camargue? **(3)**

2 a) What is meant by the 'greenhouse effect'?

b) How does the greenhouse effect lead to global warming? **(4)**

3 Explain how global warming will affect the Camargue **(5)**

4 How do environmental groups try to reduce industrial pollution in the Camargue? **(2)**

5 What action has the European Union taken to tackle global warming? **(3)**

a) List some of the problems that global warming will cause

b) Draw a poster to show the dangers brought by global warming **(5)**

38 The forests of Finland (1)

Forests under serious pressure

Before people arrived in Europe, most of the continent was covered in trees. Because of the different climates, many species of trees grew, including softwoods (such as pine) in the north, deciduous hardwoods (such as oak) in the centre and evergreen hardwoods (such as cedar) along the Mediterranean coast. Over the years these trees have been cut down to make way for farmland, settlements and roads, and to provide fuel, building material and paper. As a result, very little natural forest remains and the trees that have survived now face an extra threat—from acid rain. Those forests under the most serious environmental pressure are shown in Figure 38.1.

The forests of Finland

Finland is the most forested country in Europe. About 66% of its land is covered in trees and it depends upon them for much of its wealth. About 33% of its income and 40% of its exports are from forests and forest products. So, the country will suffer badly if its forests are destroyed and there are worrying signs that this is beginning to happen. Although some forests are cut down to create land for housing and roads, the chief culprit is industry.

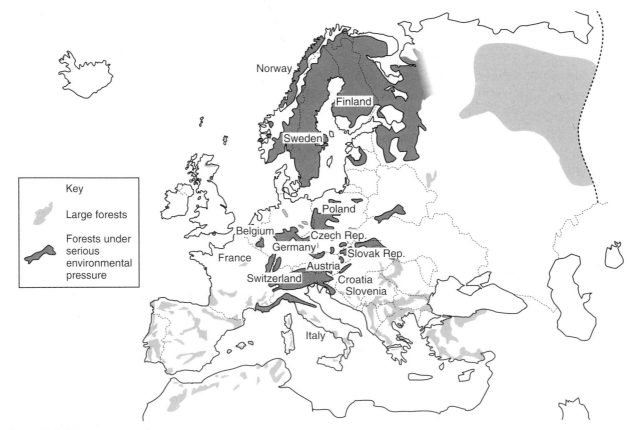

Figure 38.1 Forests under serious pressure

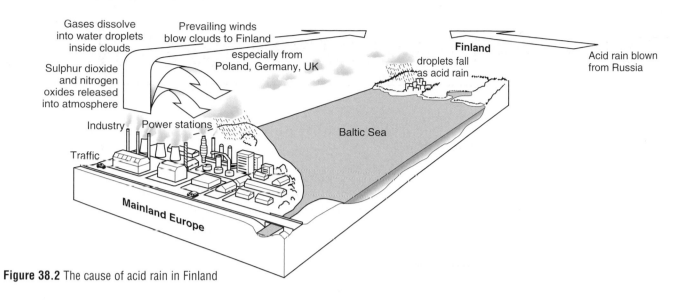

Figure 38.2 The cause of acid rain in Finland

Pressures related to industry

Acid rain

1. When fossil fuels (e.g. coal, oil, gas) are burned in power stations, motor vehicles and factories, **sulphur dioxide and nitrogen oxides are released into the atmosphere.**

2. **These gases dissolve into rainwater** to form dilute sulphuric acid and nitric acid.

3. The water droplets are carried by south-westerly winds from the industrial areas of Europe (e.g. Germany, UK) to northern European countries such as Sweden and Finland, **where they fall as acid rain.**

4. **The acid rain** seeps into the soil and dissolves important plant foods (e.g. calcium, potassium) taking them out of reach of plant roots. This process, called *leaching*, **makes the soil less fertile**.

5. **The acid rain** also releases aluminium in the soil, which **poisons tree roots**.

6. Because of leaching and poisoning, **the trees become weaker** and are more easily killed by disease, frost and drought.

Lumbering

Only 5% of **Finland's original, 'old-growth' forest** is left and some of this **is now being cut**

down to make into pulp and paper. As these trees are cut down, new forests are planted but environmentalists say that the new plantations do not have the same variety of plant and animal wildlife. Because of this, **many animals are at risk**, such as the flying squirrel, owl, golden eagle and woodpecker.

The Lapps or Saami people live by herding reindeer, trapping and berry picking within the old forests of northern Finland. As these forests are cut down, their **way of life is under threat**.

QUESTIONS

1. Name four forests in Europe under environmental pressure (2)

2. a) What are the main acids in acid rain? (1)
   b) How are these acids released into the atmosphere? (2)

3. Why does northern Europe receive so much acid rain? (2)

4. Describe the effects of acid rain on trees (4)

5. In what ways does lumbering affect **a)** wildlife and **b)** the Saami people? (6)

6* **In your own words, describe what Figure 38.2 shows** (5)

39 The forests of Finland (2)

Solutions to industry pressures

European Union countries have made agreements to reduce the amount of sulphur dioxide they release into the air. For example, by 2003 the UK will cut sulphur dioxide emissions to 60% of its 1980 level. They also have **agreements that all cars be fitted with catalytic converters**. These reduce nitrogen oxide emissions by 90%.

The **European Union** also gives **grants** to eastern European countries to reduce their air pollution. Finland is helping Estonia to clean up its air, which should mean that Finland receives less acid rain.

Environmental groups, such as Greenpeace, believe in more direct action. For example, in 1996, Greenpeace activists blocked the entrance to a pulp mill in Finland to stop old trees from being used in this way. **Friends of the Earth also name and shame logging companies** that are still felling these old trees, in their publications and on the Internet.

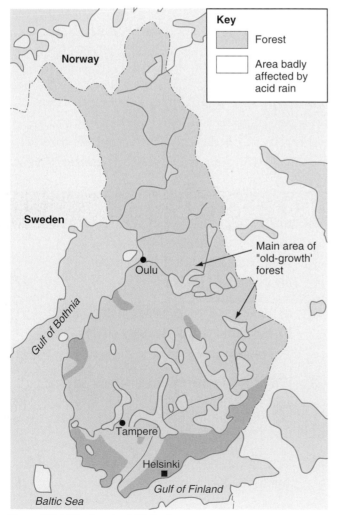

Figure 39.1 The forests of Finland

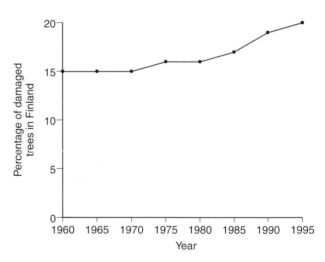

Figure 39.2 Damaged trees in Finland (1960–1995)

year	amount of sulphur (million tonnes)
1960	15
1965	18
1970	20
1975	25
1980	24
1985	23
1990	22
1995	21

Figure 39.3 Sulphur emissions in Europe (1960–1995)

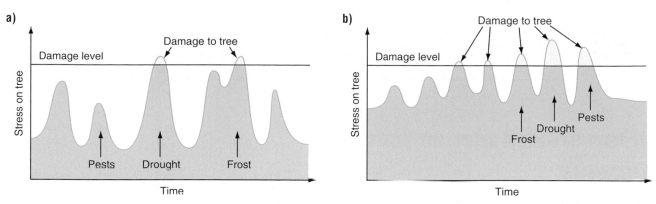

Figure 39.4 Effects of acid rain on trees a) tree stress under natural conditions b) tree stress under conditions of acid rain

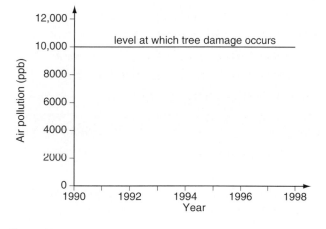

Figure 39.5 Total air pollution over south Finland (1990–1998)

year	air pollution (in parts per billion (ppb))
1990	3000
1991	4000
1992	7000
1993	4000
1994	4000
1995	4000
1996	12 000
1997	7000
1998	6000

Figure 39.6 Total air pollution over south Finland (1990–1998)

QUESTIONS

1 Describe two methods by which the European Union reduces acid rain **(4)**

2 Describe two methods by which environmental groups stop old trees from being felled **(4)**

Technique Questions

3 Describe the change in damaged trees in Finland, shown in Figure 39.2 **(2)**

4 Look at Figure 39.4.
Describe the changes in stress level in trees under a) natural conditions, b) acid rain conditions **(4)**

5 Draw a line graph to show the information given in Figure 39.3 **(3)**

6 a) Draw the line graph shown in Figure 39.5.
b) On this graph, plot the points from the table in Figure 39.6 and join them up **(3)**
c) Describe what your graph now shows **(3)**

40 Randstad, Netherlands (1)

Urban areas under serious pressure

Cities are popular places—they are popular places in which to live, work, shop and go for entertainment. But as they become popular, so their problems seem to grow as well. They find it difficult to cope with the increasing number of people and cars. This results in cities having a range of environmental pressures, and the biggest cities have the biggest pressures (see Figure 40.1).

Randstad, Netherlands

The biggest cities in the Netherlands are located around the edge of an area of fertile agricultural land. These cities have expanded until they joined together to form one conurbation in the shape of a horseshoe (see Figure 40.3). The conurbation is called *Randstad* and the farming area within it is called the *Green Heart*.

Six million people live in Randstad, making it the fourth largest urban area in Europe. It has 43% of all the Dutch people on only 17% of the land area. It does not have a single centre, but four main centres—Rotterdam (the biggest port), Amsterdam (the capital), the Hague (the seat of government) and Utrecht (the business centre). As Randstad has grown, the environmental pressures on both the urban area and the countryside have increased rapidly.

City	Population
1. Amsterdam (Randstad)	731,200
2. Rotterdam (Randstad)	593,321
3. The Hague (Randstad)	440,900
4. Utrecht (Randstad)	234,323
5. Eindhoven	148,772
6. Arnhem	135,621

Figure 40.2 Largest cities in The Netherlands

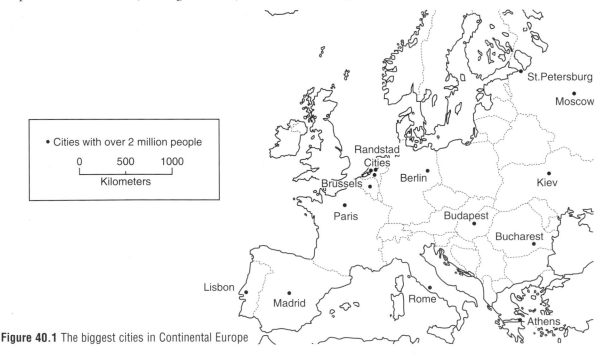

Figure 40.1 The biggest cities in Continental Europe

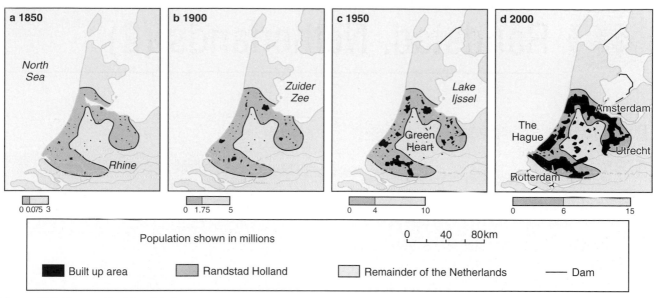

Figure 40.3 The growth of Randstad

Pressures related to urban growth

The need for more land

Figure 40.3 shows how **Randstad has expanded out into the countryside**. This urban sprawl has increased in recent years. People are moving out from the inner city areas because of the air and noise pollution, the high crime rate and the lack of open space. The population here has fallen by 15% in the last 10 years.

Shops, offices and industries are also moving out from the centre of Randstad because of the traffic congestion, the high land prices and the lack of space. They are moving to the edge of Randstad where new roads and airports are located. The number of industries and offices in the suburbs has tripled in the last 10 years.

The large population here also needs sewage works, reservoirs, and landfill sites, which are all located at the edge of the built-up area.

In total, over 50 000 hectares of the Green Heart have been built on in the last 10 years and **much of it was valuable farmland. Wildlife**

habitats have been destroyed and precious **recreational areas lost**. People living in the centre of Randstad now have much further to travel to reach the countryside. And, as urban sprawl continues, it takes the urban problems with it. The **price of land and houses goes up and traffic and pollution levels also increase.**

QUESTIONS

1. Name four urban areas in Europe under serious environmental pressure **(2)**

2. Name the four largest cities in Randstad **(2)**

3. Where is the 'Green Heart'?

4. Explain why more building land is needed around Randstad **(6)**

5. Describe two problems caused by urban sprawl around Randstad **(4)**

6* Figure 40.3 shows how Randstad has grown over the years
   a) Draw another map to show how large you think Randstad will be in the year 2025
   b) Label your map to show some problems that might occur in the future **(5)**

41 Randstad, Netherlands (2)

Pressures related to urban growth

Traffic congestion

Like most urban areas, Randstad has been settled for hundreds of years. The streets of central Amsterdam, for example, were laid out in the 17th century. Because they are so old, **the streets are narrow and have many intersections**. There are also **many narrow bridges** over the canals. This street pattern does not allow traffic to move very fast and the **traffic has also increased greatly in recent years**. This is because car ownership has risen and most office workers, shoppers and tourists now travel into the city by car.

Car exhausts increase air pollution, which affects people's health. Carbon monoxide increases pressure on the heart. The nitrous oxides increase the chances of bronchitis and pneumonia, whereas other gases can cause cancer. The effects of these pollutants are worse in the centre of Rotterdam, where the tall skyscrapers make it difficult for the pollutants to escape.

Traffic also increases noise pollution and the vibrations from buses and lorries are shaking some of the oldest buildings in the centre of Amsterdam, making them unsafe.

Solutions to urban problems

The European Union has given grants to the Netherlands to regenerate its inner city areas. By making the inner cities more attractive places in which to live, there should be less urban

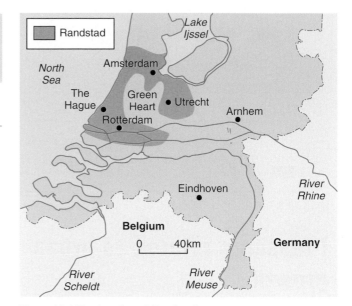

Figure 41.1 The location of Randstad

sprawl. So the European Union paid for the building of the Erasmus Bridge and a new railway station in Rotterdam and it helped to set up more cycle lanes and a 'park and ride' scheme in the city. These have helped to reduce traffic congestion. There have also been grants for new housing and for landscaping in the inner city.

Green belts have also been set up between **Randstad and nearby villages and towns**. Within these zones, very little building is allowed and this should stop Randstad from sprawling out and swallowing up other settlements.

Environmental groups, such as Friends of the Earth, are also trying to reduce these urban problems. They **have appealed to people to take part in direct action** against new road and housing schemes by

(a) writing to their politicians
(b) forming local campaign groups
(c) involving the media
(d) taking part in demonstrations

If more people do these things, Friends of the Earth believe that fewer schemes will be allowed.

Pressures related to agriculture

The 'Green Heart' in the middle of Randstad is the most fertile agricultural area in the Netherlands and is mostly used for market gardening, dairying and growing flowers and bulbs. About 40% of Netherland's farm income comes from this region. Most of it has been reclaimed from the sea (polders) and so it is below sea-level.

The biggest problem affecting the Green Heart is the **loss of farmland to urban sprawl** (described on page 81), but it has other pressures as well.

Landscape changes

Over the last 20 years, farming in Europe has become much more profitable. The prices of cereal crops, fruit, vegetables and flowers, especially, have increased rapidly. Because of this, **farmers in the Green Heart have cut down trees, drained lakes and reclaimed peat meadows in order to make more farmland**. Now, over 70% of the total area is used for farming.

But, woods, lakes and meadows provide habitats for animals and birds, which have now been lost. The wildlife has also suffered because of the **increased pollution from farm chemicals** and now many species are in decline, e.g. partridges have declined by 53% in the last 15 years and skylarks by 31%. In addition, **the landscape is less attractive** (a) because the fields are larger and there are fewer trees and ponds to provide variety, and (b) because there are now 50% more glasshouses in this area than 20 years ago.

Settlement changes

Settlements in the Green Heart are also changing.

Figure 41.2 The rise in market gardening in Holland is at the expense of the traditional landscape

There are fewer farming villages now, partly because fewer people work in farming as the use of machinery has increased. Also, as explained on page 81, people are moving out of Randstad to live in the countryside and the **old farming villages have grown into commuter settlements**.

About 700,000 people now live in the Green Heart and 15% of them are commuters. This has brought extra traffic and air pollution to the area, and the character, attractiveness and community spirit of the old villages has been lost.

QUESTIONS

1. Explain why there is so much traffic congestion in Randstad **(4)**

2. In what ways does traffic congestion affect people's health? **(3)**

3. Explain how the European Union helps Randstad to reduce urban sprawl **(6)**

4. Give four examples of direct action against urban sprawl **(2)**

5. Why are there now fewer wildlife habitats in the Green Heart? **(4)**

6. What problems have resulted from the growth of commuter villages? **(3)**

List ten facts about the Green Heart of the Netherlands **(6)**

42 Randstad, Netherlands (3)

Solutions to agricultural pressures

Since 1992 the **European Union** has changed its 'Common Agricultural Policy'. **Crop prices have been reduced** and **set-aside schemes introduced**, which should mean that fewer wildlife habitats are lost to farmland.

Growth towns have also been set up. These are new settlements in the countryside planned to take people who want to move out of Randstad. If people move here, it should prevent other villages becoming commuter towns. The growth towns should attract industries and jobs, which means that people will not have to commute into Randstad and so the amount of traffic and pollution should decrease.

Environmental groups, such as Greenpeace and World Wildlife Fund for Nature, have tried to **inform and educate the public** about the loss of wildlife habitats to farmland and cities. They research and compile information, issue press releases, publish booklets and their representatives give lectures and take part in public debates. In these ways, they help to change people's opinions so that people will not allow developments that cause environmental problems to take place.

Statistics on Randstad

area	average cost of family house (2000)
Amsterdam	£100,000
Rotterdam	£94,000
The Hague	£95,000
Utrecht	£90,000
Netherlands (average)	£81,000

Figure 42.2 Cost of housing in The Netherlands

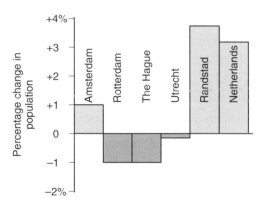

Figure 42.3 Population change in Netherlands (1994–1999)

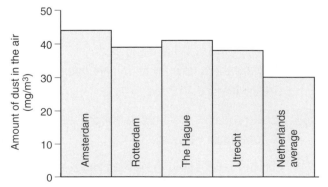

Figure 42.1 Air quality in The Netherlands

region of Netherlands	change in area of arable land (%)
the north	−2
the south	+3
the east	+5
the west (including the *Green Heart*)	−2
Netherlands (average)	+1

Figure 42.4 Changes in arable land in Netherlands (1995–1999)

QUESTIONS

1. Explain how the European Union protects wildlife in the Green Heart **(4)**

2. Describe how environmental groups can inform the public of important current issues **(4)**

Technique Questions

3. Describe the differences in air quality shown in Figure 42.1 **(3)**

4. Describe the changes in population shown in Figure 42.3 **(3)**

5. Draw a bar graph to show the information given in Figure 42.2 **(3)**

6*. **Draw a bar graph (similar to Figure 42.3) to show the information in Figure 42.4** **(3)**

Summary of environmental pressures

Pressures related to tourism
water pollution from sewage
water extraction
visual pollution
deforestation

Pressures related to agricultural areas
pollution from traffic
changes to the farming landscape
changes to settlements
pollution from agricultural chemicals
drainage schemes

Pressures related to urban areas
water pollution from sewage
pollution from landfill sites
urban sprawl
traffic congestion

Pressures related to industry
water pollution from chemicals and metals
oil pollution
thermal pollution
land and visual pollution
acid rain
deforestation

Solutions to environmental pressures

The European Union

The European Union has four main approaches:
1. **making laws/directives**, e.g. banning the dumping of metals, banning the dumping of raw sewage, forcing industries to treat their waste
2. **giving grants and loans**, e.g. for farm improvements, to regenerate inner cities
3. **issuing advice**, e.g. through environmental impact assessments
4. **having agreements between member countries**, e.g. to reduce sulphur dioxide emissions, to reduce noise pollution

Environmental pressure groups

Environmental pressure groups have five main approaches:
1. **direct action**, e.g. demonstrations against new roads and housing schemes
2. **lobbying** the people in power, e.g. governments, European ministers
3. **inform and educate**, e.g. loss of wildlife habitats, effects of global warming
4. **campaigns**, e.g. against landfill sites, for sustainable tourism
5. **name and shame**, e.g. resorts with polluted beaches, companies cutting down rare trees

43 The population of Europe

The cities of Europe

Figure 43.1 shows the location of the major cities in continental Europe. Most are found in the western half of Europe. Many are located near the coast and most are found between latitudes 45°N and 60°N. All these cities have a population of at least 250 000 and the largest, Paris, has 9 million people, but they are not particularly large when compared with cities in other parts of the world. Of the 20 largest cities in the world, only three are found in Europe (Paris, Moscow and London).

Although Europe has very few mega-cities, most of its people do choose to live in large towns and cities. **About 75% of Europe's people live in urban areas whereas in the rest of the world the figure is only 45%**. Figure 43.1 also shows how the percentage of people living in urban areas varies within Europe.

Cities are attractive places to live for many people because most jobs are to be found there, especially the highly paid jobs. There is also a wide range of services and amenities available for people to use and enjoy. **In a few European countries, such as Portugal, fewer people live in cities**. These are relatively poor countries and less industrialized, so far more people work in farming and live in the countryside.

Europe's population structure

Population structure is the make-up of the population—the number of males, females, young, middle-aged and old people. In Europe the number of males and females are very similar, except in the older age-groups. Because women live longer than men, there are twice as many women than men aged 75 and over.

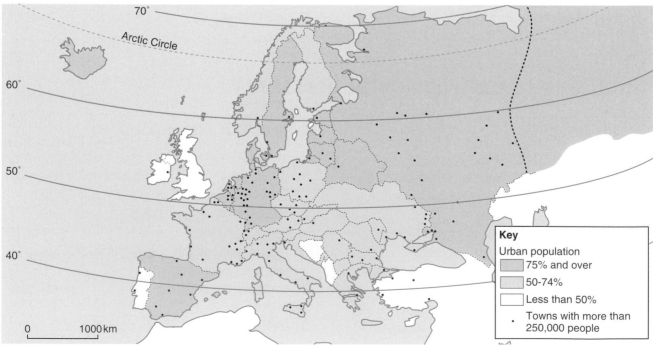

Figure 43.1 Europe's urban population

In Europe, 18% of the people are less than 15 years old. This compares with 32% for the whole world. There are fewer children because most people in Europe choose to have small families.

Some European countries have a slightly higher percentage of children (see Figure 43.2): these are countries, such as Albania, in which many people still work on the land. In farming communities, children can help on the farm from an early age and so the birth-rate tends to be a little higher here (see Chapter 45).

In Europe, 20% of the people are also aged 60 or over. This compares with only 9% in the whole world. There are more senior citizens in Europe because people live longer because of a better diet and a high standard of health care. Some European countries do have fewer old people. These are the poorer countries, such as Romania, where the health care is poorer.

country	% aged under 15	% in agriculture	% aged 60+	GNP per person (£)
Albania	33	56	8	700
Belarus	23	20	17	2000
France	20	6	19	25 000
Italy	17	9	20	19 000
Portugal	21	17	18	10 000
Romania	23	29	16	1500
Sweden	17	3	23	24 000
Switzerland	16	6	20	41 000

Figure 43.2 Population characteristics of selected European countries

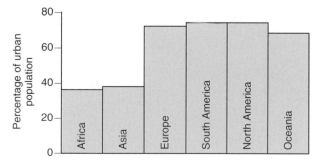

Figure 43.3 The world's urban population

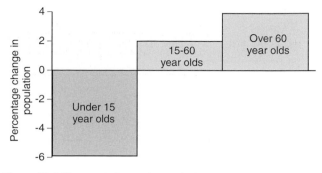

Figure 43.4 Changes in Europe's population structure (1960–1995)

population living in:	
rural areas	−17%
towns	+6%
cities	+11%

Figure 43.5 Population changes in Europe (1960–2000)

QUESTIONS

1 Using Figure 43.1, describe the distribution of cities within Europe **(3)**

2 Compare the urban population in Europe with the rest of the world **(2)**

3 Compare the percentage of people aged 15 and under and 60 and over in Europe with the rest of the world **(4)**

4 Look at Figure 43.2. For the eight countries shown, what is the connection between:
a) the percentage of people under 15 years and the percentage of people working in agriculture? **(2)**
b) the percentage of people aged 60 and over and GNP per person **(2)**

Technique Questions

5 Describe what is shown in Figure 43.3 **(3)**

6 Describe the changes in population structure shown in Figure 43.4 **(3)**

7 Draw a bar graph to show the GNP per person for the countries shown in Figure 43.2 **(3)**

Draw a graph, similar to Figure 43.4, to show the information in Figure 43.5 **(3)**

44 The distribution of population in Europe

The population of Europe is approximately 600 million and it has an average population density of 100 per km². Figure 44.1 shows the distribution of Europe's population. It is clear from this map that there are both crowded areas (**high population density**) and empty areas (**low population density**).

Several physical and human factors help to explain why some parts are more popular in which to live than others.

The importance of climate

Europe is lucky that it does not have the extremes of climate found in other continents. None of the hottest (Africa), coldest (Antarctica), wettest (Asia) and driest (South America) areas in the world are found here. Nevertheless, **there** **are still places where the climate presents so many problems that few people choose to live there.**

In northern Russia, northern Scandinavia and Iceland **the climate is particularly cold**. Few people live here because:

- the living conditions are unpleasant and also expensive, because of the high heating bills and remoteness
- the growing season is too short for crops to grow, so all food has to be imported
- it is difficult to build on ground frozen in winter, but very muddy in summer
- transport by road, rail, water and air is much more difficult in snowy conditions, so these areas can be cut off in winter
- being so remote makes them unattractive locations for industries, so unemployment is high.

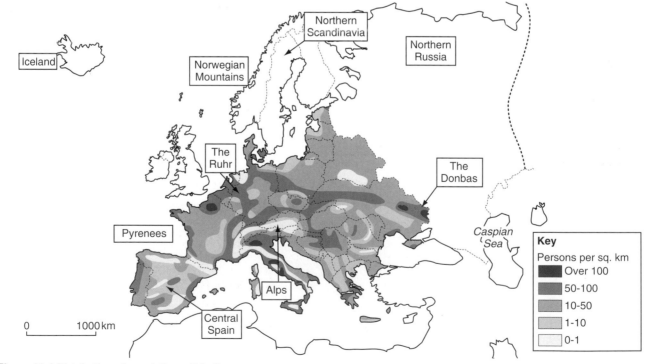

Figure 44.1 Distribution of population within Europe

Central Spain and the Caspian Sea region also have a low population density, because:

- **very little rain falls** in summer, meaning that few crops can be grown
- **soil is eroded by the wind in summer when it is dry and by rain in winter**, so farming is not very profitable
- **the temperatures are extreme**—cold in winter, very hot in summer

The importance of relief

The highest mountains of Europe—the Alps, Pyrenees and Norwegian mountains—**all have areas of low population density**. This is partly because **they are so cold**, but also because **their slopes are so steep**. This makes it difficult to build houses and also roads and railways, which makes them quite remote and unattractive to modern industry.

The importance of resources

Where the environment provides useful resources, the population density is higher. Amongst the most common resources are minerals. Wherever they are found, there are a greater number of employment opportunities. The presence of **coal**, in particular, helps to explain the high population density in areas such as the Ruhr of Germany, north-east France, Belgium and the Donbas in the Ukraine. As coal was the main source of power in the 19th century, coalfields attracted every type of industry, and their populations grew rapidly.

Other natural resources include **sandy beaches** (e.g. south of France), **beautiful scenery** (e.g. the Black Forest of Germany) and **snowy mountains** (e.g. the Alps), which become tourist

resorts and so people go to work, or, even retire there.

Other factors

The single **most important reason why any area has a high or low population density is the number of jobs available**. The area may have many employment opportunities because of its climate or because of its flat relief or because of its minerals. But there are other reasons, e.g.

- ★ areas with **busy ports** attract large numbers of people to work, e.g. Rotterdam, Antwerp
- ★ **areas with an excellent communication network** attract light industries, e.g. Frankfurt
- ★ **areas that receive government help**, e.g. grants to companies, setting up new towns, become more popular in which to live.

QUESTIONS

1. Name two areas of Europe with **a)** a high population density, and **b)** a low population density (2)

2. Explain why cold regions attract few people to live (4)

3. In what ways does relief affect population density? (4)

4. **a)** Name two coalfields in Europe. **b)** Why are coalfields densely populated? (3)

5. Name two other natural resources, apart from minerals, that attract people to live in an area

6. In what ways do governments affect population density? (2)

7*. **Make a list of all the reasons why some areas of Europe have few people and make another list of all the reasons why some areas have many people** (5)

45 Population change in Europe

Europe has a population of 600 million, which is growing at a rate of 1 per thousand (or 0.1%) per year. This is **a very slow rate of growth compared with other regions of the world**. The world average is 15 per thousand.

The population change in any region depends upon the number of births and deaths and the number of immigrants and emigrants. In a large area such as Europe, it is birth-rates and death-rates that chiefly affect population growth.

Birth-rates and death rates

The birth-rate is worked out as the number of births in a year for every 1000 people in that country. **The average birth-rate in Europe is 1·2%. This compares with 2·4% for the world as a whole**. The main reasons for the low birth-rate are shown in the star diagram below.

Death-rates are worked out in the same way as birth-rates. The average death-rate in Europe is 11. The reasons for the low death-rate are given on the next page. World death-rates are also low, but **the average life expectancy in the world is only 66 years, whereas in Europe it is 73 years.**

Subtracting the death-rate from the birth-rate gives the natural increase in population. The average natural increase in Europe is 1.2% − 1.1% = 0.1%

Europe's population has not always grown slowly. **In the past, birth-rates were much higher** because children were security for parents' old age, they were useful from an early age on farms and many children died before maturity. **Death-rates were also much higher** when health facilities were poorer.

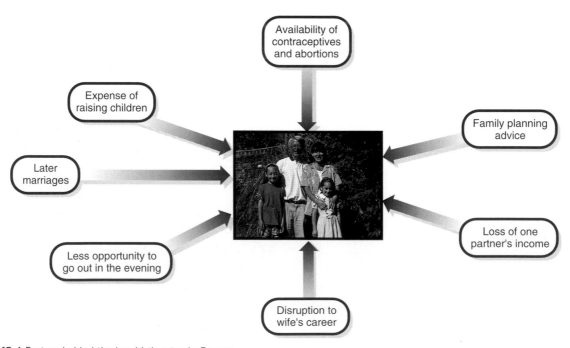

Figure 45.1 Factors behind the low birth-rates in Europe

Figure 45.2 Factors behind the low death-rates in Europe

Differences in birth and death-rates across Europe

Not every country in Europe has a very low birth-rate and death-rate (see Figure 45.3). **Some countries have a higher birth-rate and, therefore, a higher natural increase**, e.g. Albania. These are the less industrialized countries in which children are still useful on farms and there are few career opportunities for women.

Death-rates are also slightly higher in those countries that are poorer and so cannot afford

the same quality of health services, as elsewhere in Europe, e.g. Ukraine.

In countries such as Romania, the birth-rate has continued to fall until it is now below the death-rate, and their population is decreasing. This causes problems for the countries concerned, as is discussed on the next page.

country	birth-rate	death-rate	natural increase
Albania	22	8	+14
Belarus	11	13	−2
France	13	9	+4
Italy	10	10	0
Latvia	12	15	−3
Moldova	17	12	+5
Portugal	11	10	+1
Romania	10	12	−2
Sweden	11	11	0
Switzerland	11	10	+1
Ukraine	12	15	−3

Figure 45.3

QUESTIONS

1. Compare birth-rates and death-rates in Europe with those in the rest of the world **(2)**

2. Take two of the factors shown in Figure 45.1 and explain how they affect birth-rates in Europe **(4)**

3. Take two of the factors shown in Figure 45.2 and explain how they affect death-rates in Europe **(4)**

4. Give two reasons why some European countries have a higher birth-rate than others **(2)**

5. **What, do you think, are the most important reasons why**
   a) people in Europe have few children
   b) most people in Europe live to an old age?
   Give reasons for your answers **(5)**

46 Population problems and policies in Europe

Most countries in Europe have populations that are stable or growing slowly. This might seem to be an advantage for these countries. They are not going to become overcrowded and there will not be large numbers of people chasing few jobs. They do not need to build many houses, nor employ as many teachers. However, far from being an advantage, these countries regard their slowly-growing populations as a problem that they must solve.

Problems of a slowly-growing population

The problem with a slowly-growing population is its structure—in particular, **the large number of old people** (*dependent population*) compared with the **decreasing number of people of working age** (*active population*).

Because the death-rate is low and most people are living to an old age, there are an increasing number of old people in the population. This means a lot of extra expense for governments. Firstly, more pensions need to be provided. Secondly, old people become have difficulty in walking, hearing and seeing. They require many types of specialist help and equipment, they need special sheltered housing and they make up most of the hospital patients.

In contrast, because the birth-rate has been decreasing for many years, not only are there few young people, there are now fewer adults of working age. So, the country has to spend more money on providing for its senior citizens, but it has fewer working adults to pay taxes and fill jobs that create the country's wealth.

Solutions to a slowly-growing population

Countries all over Europe are looking for ways of solving the problems caused by slowly-growing populations.

In Finland, where nearly half the couples have only one child, there is **paid paternity and maternity leave**. Mothers are also given a **child-care allowance** to encourage them to return to work after their child is born.

In Sweden, because there are fewer men of working age, employers are encouraging more women to work by providing **day-care facilities** for all children aged 18 months and older. To try and increase birth-rates, **mothers are given 18 months of paid leave** when they have a baby.

In Switzerland, many **jobs are filled by guestworkers** from southern Europe and Africa.

In Greece and many other countries, to increase the Armed Forces, there is **compulsory National Service** for 18–24 months.

In Germany, because the government may not be able to provide everyone with an adequate pension, people are encouraged to take out **private pensions**. The **retirement age has also risen** in recent years.

Changes in population in Italy

Italy has the fourth largest population in Europe. Its population trends are typical of

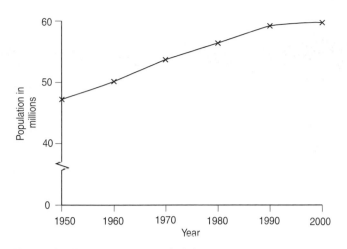

Figure 46.1 Population changes in Italy

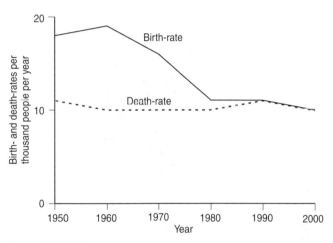

Figure 46.2 Birth- and death-rates in Italy

year	life expectancy
1950	64 years
1960	66 years
1970	70 years
1980	72 years
1990	75 years
2000	78 years

Figure 46.3 Life expectancy in Italy (1950–2000)

those in other European countries. The graphs and statistics show some of the recent changes.

year	the percentage of people who are:		
	under 15 years	15–59 years	60 years and over
1950	26	62	12
1960	26	61	13
1970	25	61	14
1980	23	61	16
1990	20	62	18
2000	17	63	20

Figure 46.4 The population structure of Italy (1950–2000)

QUESTIONS

1. What are the advantages of a slowly-growing population? **(3)**

2. Explain why there are many elderly people in a country with a slowly-growing population **(2)**

3. Describe the costs in looking after elderly people **(4)**

4. Choose two European countries and describe and explain their population policies **(6)**

5. Which, do you think, is the best way of increasing the number of people in a country: (1) increasing maternity leave, (2) attracting immigrants, or (3) raising the age of retirement?
   Give reasons for your answer **(5)**

Technique Questions

6. Describe what is shown by the line graph in Figure 46.1 **(3)**

7. Describe the changes in birth-rate and death-rate in Italy, shown by Figure 46.2 **(4)**

8. Draw a line graph to show the information in Figure 46.3 **(3)**

9* **Draw a multiple line graph (similar to Figure 46.2) to show the information in Figure 46.4** **(4)**

47   Migration movements

Migration into Europe

Most people who migrate to Europe are voluntary migrants. They *want to* leave their country, usually for a higher standard of living (**economic migrants**). In recent years, for example, one and a half million people have migrated to France from the north African countries of Morocco, Algeria and Tunisia. They were once French Colonies so many people there speak French and have links with France. France is also attractive to them because the average wage is 10 times higher, and there are more social services (e.g. doctors, hospitals, universities) and amenities (e.g. shops, entertainments).

Smaller numbers of people moving to Europe are forced migrants (called **refugees**). They move because they *have to*—because of natural disasters or war or fear of persecution in their home country.

Migration within Europe

There is a much greater movement of people within Europe than into Europe. This is partly because distances are shorter but also because the European Union allows people to work in any of the member countries. As a result, there are **many short-term migrants** or **guestworkers**, who are only working in another country for a short time. There is also **long-term migration**, and **some forced migration**, described below.

Economic migrants from Eastern Europe

Between 1945 and 1989 the countries of Eastern Europe were part of the Soviet Empire and very few people were allowed to emigrate. But, in 1989, the Soviet Empire began to crumble and people in East Germany, Poland, Czechoslovakia, Hungary, Romania and Bulgaria were then free to move to

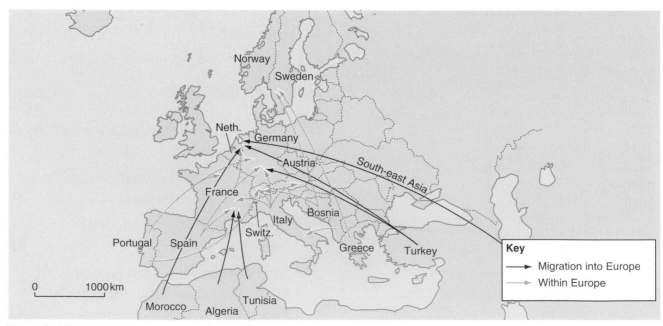

Figure 47.1 Major migration movements since 1945

other countries. Most were economic migrants. Average wages in West Germany, for example, were five times those in Eastern Europe. In 1990, East Germany and West Germany became one country once again. Over the next 5 years, 100 000 people from East Berlin found new jobs in West Berlin.

Figure 47.2 East Europe in 1989

This movement of people from Eastern Europe into the new Germany caused serious problems. Unemployment rose, there were housing shortages and taxes had to be raised, so lowering people's standard of living. There was social unrest, with many strikes and demonstrations. There are problems also for the countries of the old Eastern Europe. The loss of young adults is causing labour shortages there, slowing down their economic development.

Forced migration from former Yugoslavia

Yugoslavia used to be a nation of six regions, but it broke up into independent countries in the early 1990s (see Figure 47.3). In 1992, **Serbia and Croatia went to war**, as each was trying to take over Bosnia and many Bosnian Muslims were **forced to flee to escape ethnic cleansing**. Then, in 1998, Serbs began to force out Albanians from the province of Kosovo in

Serbia. Many were forced to flee to Montenegro. The weather was very cold, they had no shelter, many became ill and some even died. They set up camps on the border until they were able to return under the protection of NATO forces in 1999.

Figure 47.3 East Europe in the year 2000

QUESTIONS

1 What is **(a)** an economic migrant, **(b)** a refugee, and **(c)** a guestworker? **(3)**

2 Explain why people migrate from North Africa to France **(4)**

3 **a)** Explain why so many people have moved recently from Eastern Europe to Germany **(6)**
b) Describe the effects of this migration on Germany and the countries of Eastern Europe **(6)**

4 Explain why people have been forced to emigrate from the former Yugoslavia **(3)**

Look at Figure 47.1. Name the European countries that people have migrated *from*, and those that people have migrated *to* (5)

48　Urban change in Berlin, Germany (1)

Figure 48.1 The Brandenburg gate, separating East and West Berlin, a monument to Berlin's important history

Introduction

Berlin has always been one of Europe's most important cities but after the Second World War, most of it lay in ruins and it was then split into two parts. East Berlin became capital of the new communist country of East Germany. West Berlin was still part of West Germany, but 175 kilometres from its border (Figure 48.2).

In 1989, the Berlin Wall, which separated East and West, came down and Berlin became one city again. One year later, East and West Germany became one country again (*reunification*) and, in 1999, Berlin became the capital of the unified Germany.

Changing back from two separate cities into one has caused immense problems. No city in Europe is changing more quickly. In the 1990s, Berlin became the biggest building site in Europe.

Housing changes

Berlin urgently needs more housing. This is partly because its population is rising—from 3.5 million in 1995 to an estimated 4.5 million in 2010. It is also because more people now live on their own.

Housing in the central area (e.g. Kreuzberg)

Inner city Berlin has had similar problems and policies to those found in other European cities, e.g. Glasgow, Paris. The original housing was built in the 1850s and was made up of five-storey apartments. They were overcrowded (over 1000 people per hectare), lacked amenities (e.g. baths, hot water) and, by the 1960s, some were decaying with old age. **Slum clearance took place and cheap apartment blocks were built**, but many people still had to be rehoused in new suburban schemes.

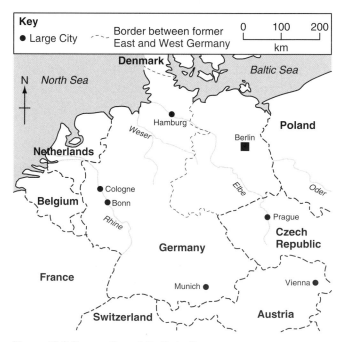

Figure 48.2 The position of Berlin in Germany

In the 1980s, many of the **older properties were renovated**, rather than being pulled down. This meant adding amenities, e.g. baths, reroofing, rewiring and general modernisation. It was cheaper than slum clearance and caused less upheaval.

Both these policies resulted in fewer people living in the inner city. Now the **Berlin authorities wish to encourage more people to live in the central area**. Any new development in the city centre now must include at least 20% of its space for housing. For example, new office blocks in the Potsdamer Platz in the heart of Berlin also have housing for 2000 people.

To build all the new housing needed will cost Berlin £7.5 billion. To renovate apartments and bring them up to standard will cost another £4.5 billion. The city also has additional problems in its inner city.

The inner city housing is the poorest and cheapest and, therefore, **attracts most of the socially disadvantaged population**, including senior citizens, single parents, the disabled and immigrants. These people require more social services, which is another expense for the city. In addition, immigrant people find it much more difficult to obtain jobs because of language difficulties or discrimination. Whereas other groups of people may find jobs, become wealthier and move away, immigrants tend to stay in the inner city. **In time, areas of the inner city become ghettoes**—a ghetto is an area of a town in which most of the people belong to one minority group. In Berlin, Turks are the biggest minority group. The people here need more help, such as language classes and job training. There is also occasional unrest and fighting between the Turks and neo-Nazi groups. It gives the authorities more problems and costs more money.

Figure 48.3 A Turkish ghetto in Berlin

QUESTIONS

1 Since 1989, Berlin has changed from being two cities to one city again. Explain what this means **(2)**

2 Describe the problems with the old houses in central Berlin **(3)**

3 Describe three solutions to the housing problems **(3)**

4 What is a ghetto and why did they develop in Berlin? **(4)**

5 In what ways do ghettoes cause the authorities problems? **(3)**

6 Apart from immigrants, what other disadvantaged groups live in central Berlin and why do so many live here? **(3)**

7* **Look at Figure 48.3**
a) Describe what the photograph shows
b) Where in Berlin will you find areas like this?
c) What problems are there in these areas? (5)

49 Urban change in Berlin, Germany (2)

Housing in the suburbs (e.g. Biesdorf)

When the slums were being cleared in the inner city in the 1960s and 1970s, **large estates of ten-storey apartment blocks were built in the suburbs**, e.g. Marzahn. They each housed 50 000 people, but had very few shops, schools and other services. They were cheaply made and were nicknamed *worker cupboards* by the residents.

Since 1990 a series of **growth centres have been built at the edge of Berlin**, e.g. Biesdorf (Figure 49.1). To attract large numbers of people, these centres not only have a variety of housing, they also contain shopping centres and recreational areas, and have business parks with many jobs available.

People are also moving beyond Berlin's boundaries to live in the Brandenburg region that surrounds it. This is **called counter-urbanisation**. They are moving there because the price of housing is cheaper, the quality of the environment is much better and there are now fast rail links into Berlin.

With so much development taking place, Berlin is expanding outwards rapidly. This urban sprawl is swallowing up separate villages and towns, as

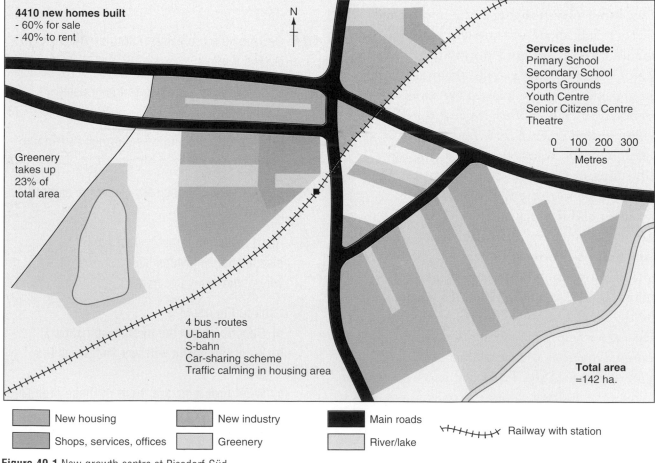

4410 new homes built
- 60% for sale
- 40% to rent

N

Greenery takes up 23% of total area

Services include:
Primary School
Secondary School
Sports Grounds
Youth Centre
Senior Citizens Centre
Theatre

0   100   200   300
Metres

4 bus -routes
U-bahn
S-bahn
Car-sharing scheme
Traffic calming in housing area

Total area =142 ha.

| | New housing | | New industry | | Main roads | | Railway with station |
| | Shops, services, offices | | Greenery | | River/lake | | |

Figure 49.1 New growth centre at Biesdorf-Süd

well as farmland, woodland and wildlife habitats. It is hoped, however, that **the new growth centres in the city will reduce urban sprawl**, because more people will want to live here rather than in the countryside.

Transport changes

Changes in the central area

During World War Two, Berlin's **roads and railways were devastated** and, while the Berlin Wall existed, there were **no transport links between the two cities**. This is because people were not allowed to travel between East and West Berlin. **The main task, therefore, has been to reconnect the dead-end streets** that were separated by the Berlin Wall. This has made it easier for road traffic to move around and also made it possible to develop a proper city bus service. **Underground train stations** (U-bahns) which had been closed in East Berlin **have been opened up** and now provide an alternative means of transport.

The city also wants to reduce traffic congestion and the air pollution it causes. Its plan is for 80% of all journeys to be on public transport and it has tried to improve the railways and buses in particular.

In 1990, **they brought in a car-sharing scheme**. People can pay £10 per month to share cars that are spread around the city. They must phone to book their car, and collect the keys from a safe deposit box nearby. Then they only have to pay for the time they use the car and the distance they travel. This should reduce the number of cars on the road.

Changes in the suburbs

In an attempt to **improve public transport**, new above-ground electrified railways (S-bahns) have been built, connecting the centre with the suburbs, and 800 kilometres of cycle ways have also been laid out in order to encourage commuters to cycle into work.

A ring-motorway now completely encircles Berlin (motorway 10) and should reduce the amount of 'through' traffic on the city's roads.

Figure 49.2 A U-bahn station in Berlin

Travelling by air is more popular, but two of Berlin's three airports cannot expand. They are too close to houses, roads and railways. So, **a new airport has been planned** for the year 2007 (to be called Berlin–Brandenburg Airport). It will be located next to Berlin's other airport in the south of the city.

QUESTIONS

1. Describe the housing schemes built in the 1960s in Berlin's suburbs **(4)**

2. Biesdorf-Süd is a new housing area. Describe the advantages of living here **(6)**

3. Explain how new growth centres, such as Biesdorf-Süd, will reduce urban sprawl **(2)**

4. When Berlin became one city again, its transport system had many problems. Describe two problems with the road system, one problem with its airports, and one problem with its railways **(4)**

5. Describe the different ways in which Berlin is trying to reduce the car traffic in the city **(6)**

    Look at Figure 49.1. It shows a new housing area in Berlin.
    What, do you think, would be good and bad about living in this area? **(5)**

50 Urban change in Berlin, Germany (3)

Shopping changes

Shopping in the central area

Because East and West Berlin were separate cities, they both had central business districts (CBD). Since 1989, **the CBD in the east has expanded rapidly**. A large attractive shopping area with many restaurants and entertainments is being built at Potsdamer Platz and this will probably become the main CBD. **The shopping centre in the west is suffering from competition** with the east and with the new shopping centres in the suburbs. Some shops may have to close and jobs will be lost. Others have adapted by becoming speciality shops, e.g. expensive clothes shops, exclusive jewellers.

Shopping in the suburbs

The 1990s saw **a rapid growth of planned suburban retail parks** in Berlin, such as at Wildau and Falkenberg (see Figure 50.1). With a mixture of shops, entertainments and restaurants, they are proving very popular. This attracts even more shops away from the CBD, increasing the problems there. They also bring a few of the CBD problems with them. Because they are so

Variety of middle-order and high order shops

Beside ring-motorway (A10)

Hotel

Fast-food restaurants

Cinema

New traffic lights - traffic congestion here

Cheap land

Garden centre

Flat land

Space for large free car parks

Greenfield site - it was farmland

Not a covered 'mall' but a retail park

Figure 50.1 New suburban shopping centre at Falkenberg

popular, traffic congestion can occur at certain times and there is extra air and noise pollution here. Also, because they have been built at the very edge of the city, they have taken over a significant amount of farmland and woodland.

Recreational changes

Berlin has more recreational land than most large cities. One-quarter of the city is woodland and, in addition, there are many parks, rivers, lakes and canals which all offer opportunities for recreation.

Changes in the central area

There are fewer recreational areas near the centre of the city because the land is so expensive and in demand for shops, offices and industry. Nevertheless, in the heart of the city, separating the two shopping centres is Tiergarten. This is a large park, stretching for two kilometres along the River Spree, and containing woodland, paths and ponds, as well as Berlin Zoo. **There has been an increase in indoor facilities**, such as fitness studios and health clubs, in the centre. They do not require much land and they are convenient for office workers and shop assistants to use at lunch-time or after work.

Changes in the suburbs

Recreational uses that take up a lot of land are found in the suburbs where the land is cheaper. Berlin's football team plays in the Olympic Stadium, built for the 1936 Olympic Games, which is in the inner suburbs. Golf is growing in popularity and **many golf courses are being built around the edge of Berlin**. There is **a new theme park** here as well, called Filmpark.

Just like the new suburban shopping centres, **these new recreational uses increase traffic congestion and urban sprawl**. The Berlin authority's solution has been to develop growth centres (see Figure 49.1). Within these housing areas are new sports halls, playing fields and play parks. With **a lot of facilities now provided near to where people live**, there should not be as much need to take over farmland at the edge of the city.

Figure 50.2 Filmpark, Berlin's new theme park.

QUESTIONS

1 What changes are taking place in shopping facilities in the centre of Berlin? **(3)**

2 **a)** Describe the characteristics of a new suburban shopping area
b) What problems are these new centres causing? **(4)**

3 What new recreational facilities are being built in Berlin and where are they being built? **(5)**

4* **Look at Figure 50.1.**
Why, do you think, the new shopping centres in the suburbs of Berlin are so popular? **(5)**

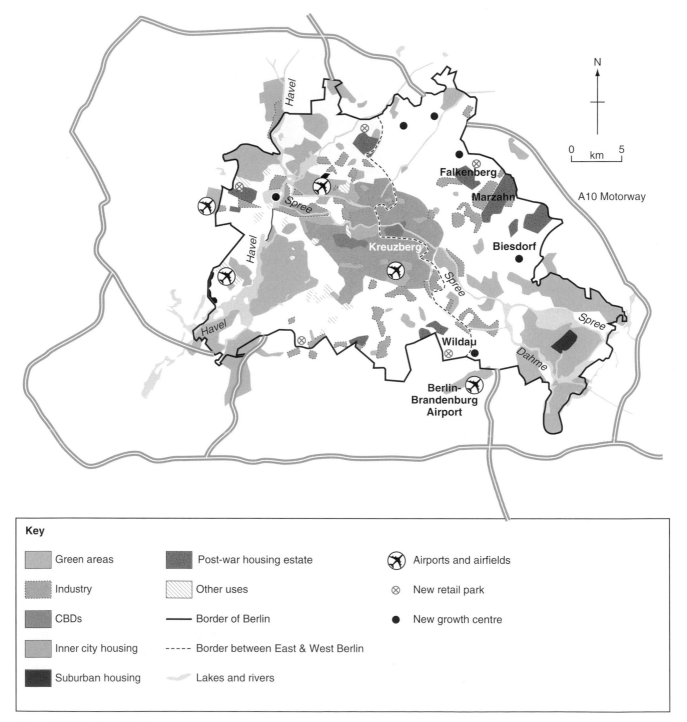

Key

Green areas	Post-war housing estate	Airports and airfields
Industry	Other uses	New retail park
CBDs	—— Border of Berlin	New growth centre
Inner city housing	- - - - Border between East & West Berlin	
Suburban housing	Lakes and rivers	

Figure 50.4 The city of Berlin

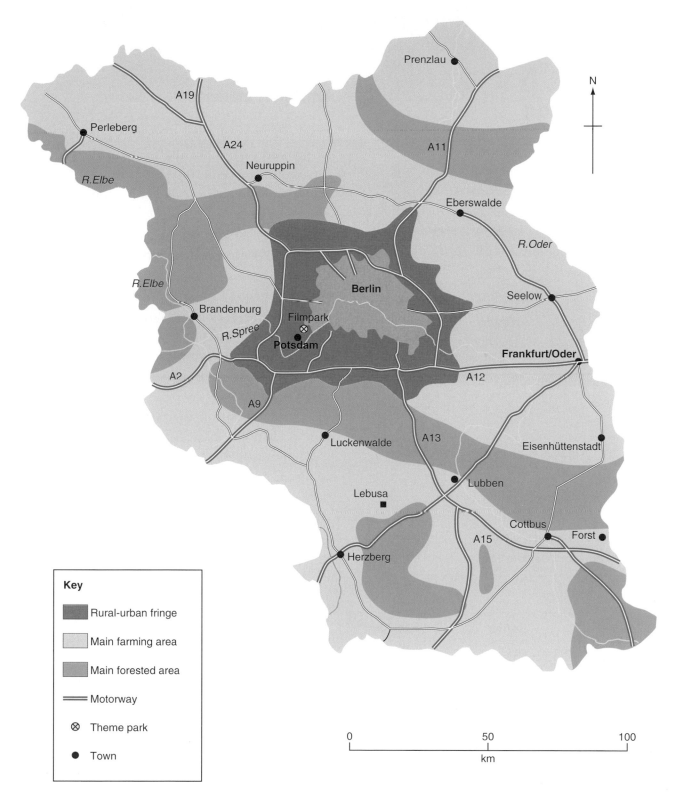

Figure 50.5 Brandenburg, Germany

51 Rural change in Brandenburg, Germany (1)

Introduction

Brandenburg is a large region in the north-east of Germany that surrounds the city of Berlin. It was part of communist East Germany until reunification with West Germany in 1990. It has been an economically backward region, devoted mostly to farming, but it looks set to benefit from the growing wealth of Berlin.

The area around Berlin is beginning to change quite rapidly as the city spreads outwards. Away from Berlin, however, the major changes taking place are in farming. Overall, Brandenburg is now a very different region from what it was in the 1980s.

Rural change near Berlin

The area of countryside next to a city is called the *urban-rural fringe*, and is an area where a lot of changes in land use take place. The urban-rural fringe around Berlin is no exception. This area is changing faster than any other part of the Brandenburg region.

characteristic	Berlin	Brandenburg
area (km²)	891	29,476
population (1997)	3.50 million	2.57 million
population density	3928	85
% under 15	14	15
% 15–64	67	64
% 65 and over	19	21
birth-rate%	0.86	0.53
death-rate%	1.22	1.15
GDP/person	£10,049	£4,611
unemployment	11.1%	15.9%

Figure 51.1 Characteristics of Berlin and Brandenburg

Transport changes

In recent years, **the road system has been improved** with more and wider dual-carriageways and motorways. Now, six motorways radiate from the city, while another one encircles the city as a giant ring-motorway. An inner ring-motorway is almost complete (see Figure 50.5). But, recent evidence suggests that building more roads does not solve congestion. It only encourages more people to travel by car. So the authorities are now discouraging car-driving. Instead, they are improving the buses and trains and creating more cycle ways (see page 99). The railways have been improved by building high-speed rail links to other cities in Germany, but these can also be controversial. The high-speed link to Hamburg caused many complaints when it was built through a habitat for rare birds. Because of this, no construction work was allowed between the months of March and September.

Land use and settlement changes

Partly because of the shortage of land in Berlin and partly because of the improved transport, **Berlin is growing outwards into the countryside**. More and more people are building houses in the countryside and travelling into Berlin to work. The fast roads and railways also attract new offices, shops, factories and recreational facilities (see Figure 51.2). For instance, in the first 5 years after unification, there were applications for 40 new golf courses and 27 new theme parks in the urban–rural fringe.

All these urban land uses take over a lot of rural land, which means fewer crops are grown and fewer animals reared. As the city sprawls out, it reaches once separate villages that then become suburbs of the city, e.g. Spandau.

To try and stop this urban sprawl, **the authorities are building new growth centres within the city** (see page 98) **and they have set up protected areas around the city**, similar to green belts. Building is restricted in these areas, which stops the city from spreading. However, Berlin has a shortage of housing and rising unemployment. This makes it very difficult for planners to refuse applications from companies who wish to build housing estates, business parks or shopping centres in the countryside.

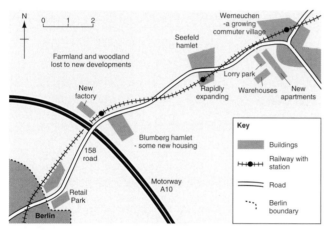

Figure 51.2 Recent developments north-east of Berlin

Settlement changes

The villages are also changing rapidly. Villages on or near main roads or railway stations are growing as commuter villages. **Houses are now**

being bought by commuters, instead of being rented to farmworkers. More remote villages are in decline (see Figure 51.3)

(see page 98)

QUESTIONS

1. Where is Brandenburg? **(2)**

2. What is the 'urban-rural fringe'? **(1)**

3. Look at Figure 51.2. Describe and explain the developments taking place north-east of Berlin **(6)**

4. Describe the problems that occur as Berlin spreads outwards **(2)**

5. Explain how the authorities can reduce these problems **(2)**

6. Explain why some villages near Berlin are growing, whereas others are declining **(3)**

7. Describe one problem facing **(a)** remote villages and **(b)** more accessible villages in Brandenburg **(4)**

8* **Look at Figure 51.1. What are the main differences between Berlin and Brandenburg** **(5)**

accessible villages	remote villages
up to 50% of working population work in Berlin	mostly farmworkers and retired people
population rising	population falling
high standard of living	low standard of living
new buildings, of different style and materials from older ones	many derelict buildings and empty houses
properly surfaced, wider roads	cobbled and poorly-surfaced roads
several shops + hairdressers, banks, cafes	few shops and services
properly landscaped gardens and parks	untidy and untended areas of greenery
house prices rising	house prices falling

Figure 51.3 Village changes in Brandenburg

52 Rural change in Brandenburg, Germany (2)

Changes in the countryside

The Brandenburg region is very much affected by the city of Berlin that it surrounds. Away from Berlin, it is mostly a farming area. The changes here have much to do with its communist past.

Farming changes

Brandenburg is a poor farming region. Its soils are mostly glacial outwash sands and terminal moraines, separated by wide, marshy valleys. Hardy crops such as rye and potatoes are grown and, in the more fertile, drained valleys, wheat and sugar beet are more common. About 35% of the whole region is forested.

When Brandenburg was part of communist East Germany, the farms were extremely large and owned by the state. Because everyone was guaranteed a job, the farms employed many people but could not afford other inputs, such as machinery. In 1989, for example, West German farms had 12 tractors for every 100 hectares of land, while East German farms only had three. Crop yields were 20% less than in West Germany. Since Brandenburg became part of the unified Germany in 1990, **the large state farms**

have been broken up. They are now much smaller (see Figure 52.2) and **many have been returned to their former owners**. Some farmworkers have bought farms, but they are too expensive for most farmers. Instead, in many areas, **groups of farmers have formed cooperatives and bought up farms.**

Number of Farms			
Farm Size (hectares)	1992	1994	% change
<1	450	204	−55%
1–50	3241	4128	+29%
50–200	688	941	+37%
200–500	220	437	+99%
500–1000	215	293	+36%
1000–2000	300	313	+4%
>2000	157	127	−19%

Figure 52.2 Changes in farm size in Brandenburg (1992–1994)

Brandenburg is now in the European Union and so its **farmers are affected by the Common Agricultural Policy**. This has already led to changes in land use. Instead of growing crops they were told to grow by the state, **farmers are now free to grow high-priced crops**, such as oilseed rape. They are also paid to take some of their arable land out of production (*set-aside land*). The European Union has given grants to farmers here so that they can use more machinery on their land. **The increased use of machinery**

Figure 52.1 East German state farm in the 1980s

Figure 52.3 Farm ownership around Lebusa in 'East' Germany, post-unification (1989)

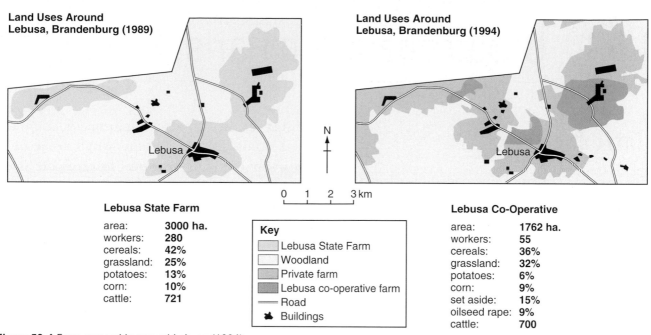

Land Uses Around Lebusa, Brandenburg (1989)

Land Uses Around Lebusa, Brandenburg (1994)

N

0 1 2 3 km

Lebusa State Farm

area:	**3000 ha.**
workers:	**280**
cereals:	**42%**
grassland:	**25%**
potatoes:	**13%**
corn:	**10%**
cattle:	**721**

Key
- Lebusa State Farm
- Woodland
- Private farm
- Lebusa co-operative farm
- Road
- Buildings

Lebusa Co-Operative

area:	**1762 ha.**
workers:	**55**
cereals:	**36%**
grassland:	**32%**
potatoes:	**6%**
corn:	**9%**
set aside:	**15%**
oilseed rape:	**9%**
cattle:	**700**

Figure 52.4 Farm ownership around Lebusa (1994)

has led to **a huge reduction in farm workers**. In the first 5 years after unification, the number of farmworkers dropped by 80%, increasing unemployment levels in Brandenburg and causing villages to decline rapidly in population.

Housing changes

Before reunification, people working on the huge state farms mostly lived in villages, such as Lebusa (Figure 52.4). The **houses were all owned by the state** but **they were rented very cheaply** to the people. Everyone could afford a place to live.

Since 1990, many of the houses have been returned to their former owners. **People are now free to buy and own houses**. Many cannot afford to do this and, instead, they rent. But the **rents are now much more expensive** because they are no longer subsidized by the state. This is another reason why people are leaving the countryside, leaving empty houses in villages. Many of the people who have bought their own farms are now **building new farmhouses in the middle of their land**, instead of living in villages.

QUESTIONS

1. Look at Figure 52.2.
   Describe the changes in farm size in Brandenburg between 1992 and 1994 **(4)**

2. Describe one problem caused by the break up of state farms **(1)**

3. Describe two changes in land use on farms since 1990 **(2)**

4. Describe and explain the changes in agricultural employment since 1990 **(3)**

5. Look at Figure 52.4.
   Which land uses have increased in Lebusa since 1990? **(3)**

6. a) In what way has home ownership changed in Brandenburg since 1990? **(2)**
   b) Describe one problem caused by this **(2)**

   Look at Figure 52.4.
   Compare the farming in the area of Lebusa in 1989 and 1994 **(5)**

53   Urban change in Paris, France (1)

Introduction

Paris grew up at an island and crossing-place of the River Seine (see Figure 54.1). So, from early times, it has been both an important route-centre and a port. As long ago as the 12th century it became the capital and its importance grew and grew. Today, with 9.5 million people it is eight times greater than any other city in France and it is the biggest city in the whole of Europe.

But, a large and growing population brings problems, and, by the 1950s, some of its problems had become quite serious. As a result, there have since been major changes to the Paris landscape.

Housing changes

After World War Two, the birth-rate in Paris rose, immigrants began to move in from other parts of France and abroad, and so the population began to grow quite rapidly. At the same time, much of the housing in the city was at least 50 years old and becoming unfit to live in. Paris, therefore, faced a critical housing shortage.

Housing changes in the central area

In the 1950s, the housing in the inner city was mostly seven- or eight-storey tenements, in which people lived in one or two rooms without running water or their own toilet. It was extremely crowded, about 40 000 people to every square kilometre.

Much development has taken place since then. **The worst slums were pulled down, while others were renovated**. Some were bought cheaply by wealthy, middle-class people who work in central Paris. They have improved the

housing themselves which, in turn, attracts other people to do likewise. In time, **small areas of the inner city can become 'fashionable' for the more wealthy**, e.g. Belleville, Bercy. **This is called gentrification** and it is one reason why more people are now moving back into the central area.

The other main reason why people are returning is because of **the new regeneration schemes. These schemes do not just improve the housing. They also improve the environment and the facilities and attract industries**. The biggest example in Paris is at La Defense.

Figure 53.1 La Defense in Paris

In the 1960s, a large area of tenements and workshops was pulled down and 25 tower blocks were erected, up to 40 storeys high. Thirty-five thousand people now live in these high-rise blocks, attracted by the improved housing, the new offices and the shopping facilities. La Defense is a traffic-free area, with all the roads and car parks located underground, as is the metro (railway) station connecting the area to the city centre.

Housing changes in the suburbs

While redevelopment was taking place in the inner city in the 1960s, **vast new housing schemes were built in the suburbs**. They were mostly high-rise flats called *grands ensembles*, for example at

Sarcelles. They housed very large numbers of people, but were unpopular as they had few shops, jobs and poor public transport. As some people moved away, immigrant families, especially from North Africa, moved in. Racism by some landlords meant that they found it difficult to live anywhere other than the poorest areas.

In addition, many immigrants preferred to live within their own ethnic community where people spoke the same language and had the same religion and customs. In this way, the percentage of ethnic minorities increased and **some of these housing schemes became ghettoes** (see page 97). Some immigrants, unable to find proper housing, built rough shacks in these areas. These shanty towns are called *bidonvilles* in France (literally, petrol can towns).

At the same time, because Paris had such a serious housing problem, **five new towns were built** up to 25 kilometres from the city, e.g. Cergy-Pontoise (see Figure 54.2). Up to 200 000 people live in each new town. **More recently 'growth centres' have been set up**, such as Roissy, outside the city. New housing and industry is encouraged to set up only in these centres, to prevent urban sprawl in other areas.

Shopping changes

The distribution of shops in Paris has changed much in the last 30 years. This is partly because people have moved out to the suburbs. It is also because more people own cars and find it easier to travel to shopping centres further away.

Shopping changes in the central area

Most of the small 'corner shops' in the old inner-city have now gone. Some were pulled down with the housing in the 1960s. Some went out of business as people moved away and those that were left travelled to cheaper supermarkets elsewhere.

As part of the regeneration of inner-city Paris,

the biggest shopping centre in Europe was opened in La Defense in 1981.

In recent years **the city centre has become less popular with shoppers**. Traffic problems put people off, especially when there are suburban shopping centres nearer to where they live. The city centre is now dominated by luxury shops, sports shops and clothes shops.

Shopping changes in the suburbs

The **last 25 years has seen a large number of hypermarkets being built** in suburban shopping centres. These shops are larger than supermarkets and sell a wider range of goods, including clothes, electrical equipment and even furniture, at very competitive prices. They have located along main roads at the edge of Paris where access is easy and land is cheap. **In the last few years, however, their growth has slowed down**. They are blamed for adding to the urban sprawl and the authorities now wish to encourage more people to shop in the city itself and will not allow the number of hypermarkets to increase.

QUESTIONS

1 Describe the problems of the old housing in Paris (3)

2 In what ways is La Defense a better area to live than it used to be? (5)

3 In what ways did suburban housing schemes fail? (3)

4 Where were extra houses for Parisians built outside of Paris? (1)

5 In which areas of Paris have shops increased and where have they decreased? (3)

6 a) What are hypermarkets? b) What problems do they cause? (4)

7* **List five housing changes and five shopping changes in Paris since the 1960s** (5)

54 Urban change in Paris, France (2)

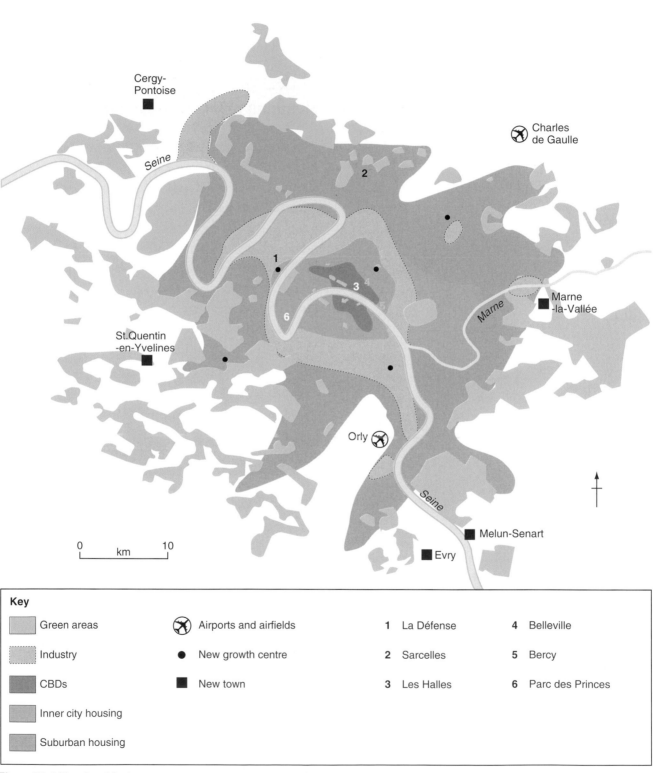

Figure 54.1 The city of Paris

Key			
Green areas	Airports and airfields	1 La Défense	4 Belleville
Industry	• New growth centre	2 Sarcelles	5 Bercy
CBDs	■ New town	3 Les Halles	6 Parc des Princes
Inner city housing			
Suburban housing			

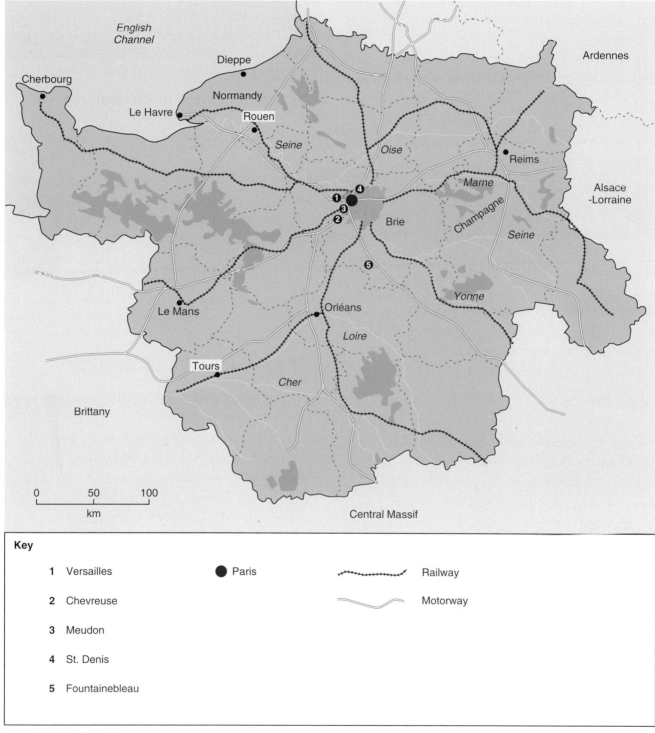

Figure 54.2 The Paris Basin

Key

1 Versailles
2 Chevreuse
3 Meudon
4 St. Denis
5 Fountainebleau

● Paris

┼┼┼┼ Railway

──── Motorway

Transport changes

Transport changes in the central area

The average speed of traffic in central Paris is 10 kmph, less than it was 100 years ago when people travelled by horse and carriage. As in most European cities, **traffic congestion is a major problem**. 1.5 million cars head for the centre of Paris every day but to get there most of the traffic has to cross the River Seine, and there are few bridges by which to cross it. Within the old core of the city, the streets are narrow, with many intersections and, overall, there are parking places for less than one million cars.

The city has tried several approaches to solve traffic congestion. In the 1960s and 1970s, **the policy was to improve conditions for the motorist**. An inner ring-road was built (the *Boulevard Peripherique*), roads were widened and underground car parks provided. At the same time, **many offices and industries were decentralized**. They moved out either to the suburbs or other towns, which reduced the number of people needing to travel into the centre.

In the 1990s it was realized that improving roads only encouraged more people to travel by car. So, **the authorities now discourage people from using cars**. New metro (underground railway) lines have been opened, bus and cycle lanes have increased and more streets have become pedestrianized, e.g. *la voie verte* (the green road). Once a month, Sundays are dedicated to cyclists and cars are banned from many areas. Also, when air pollution is bad, only cars with an even registration number are allowed into the centre on one day and only cars with odd numbers on the following day. At these times travel on the metro is free.

Figure 54.3 Pedestrianised street in the centre of Paris

Transport changes in the suburbs

Road, rail and air transport in the suburbs is improving. New metro lines have been built into the suburbs and as far away as the new towns. As well as the *peripherique*, an outer ring-motorway will be completed early in the 21st century (the A86). To relieve congestion at Paris's main airport (Orly), Charles de Gaulle airport was built in the 1970s on flat land beside the A1 motorway and away from housing areas. It now is also expanding.

Recreational changes

Recreational changes in the central area

Perhaps surprisingly, **the number of parks in the central area is increasing**. Some have been reclaimed from derelict land. For instance, one park was a refuse tip for dead horses; another is on the site of a slaughterhouse. Recent regeneration projects have also led to an increase in various recreational facilities. At Les Halles, amenities include a swimming pool and gym. These help to attract people back to live in the central area. French rugby internationals are played at the Parc des Princes, which is also the home of Paris's biggest football team—Paris St Germain. This is found at the edge of the central area, near the Bois de Boulogne.

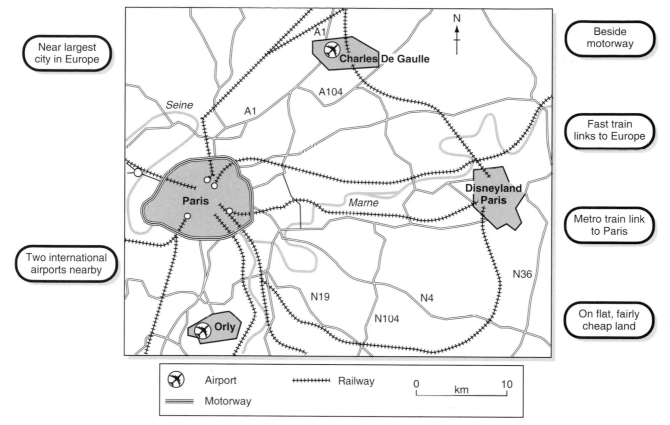

Labels on map:
- Near largest city in Europe
- Beside motorway
- Fast train links to Europe
- Metro train link to Paris
- Two international airports nearby
- On flat, fairly cheap land

Map features: Charles De Gaulle, A1, A104, Seine, Paris, Marne, Disneyland Paris, Orly, N36, N19, N4, N104

Key: Airport · Railway · Motorway · 0 km 10

Figure 54.4 Location of Eurodisney

Recreational changes in the suburbs

There are over 350 parks in Paris and most are in the suburbs. At the edge of Paris, however, **traditional recreational areas of forests, parkland and water have been lost** to urban sprawl. So, **a 'green belt' was set up**, which has preserved areas of recreation in many areas of the city, such as the Bois de Boulogne. To compensate for the loss of green areas, more parks are being set up nearer the centre, as described above. Also, in 1992, another type of recreational area was opened near Paris. **Eurodisney theme park was built** on farmland just east of Paris. Figure 54.4 explains why it was set up here. It attracts millions of people as well as providing recreation for Parisians. In addition, it has created 14 000 jobs, many more tourists and, consequently, much more money into the city.

QUESTIONS

1 Describe ways in which the Paris authorities try to reduce the number of cars in the centre **(4)**

2 **a)** Where are recreational areas in Paris being lost?
   b) Describe different ways in which this problem is being solved **(5)**

3 Describe and explain the location of Eurodisney **(6)**

4 Suggest **a)** developments that are likely to take place next to Eurodisney, and **b)** problems that might be caused by locating Eurodisney here **(6)**

5 **a) Give reasons why Disney chose to build a theme park near Paris**
   b) What problems do you think this might have caused? (5)

55 Rural change in the Paris Basin, France (1)

Introduction

As the name suggests, the Paris Basin is a lowland area surrounding the city of Paris. Although low land it is rarely flat, being more a mixture of ridges and valleys drained by the River Seine and its tributaries. It is the most important agricultural region in France, containing such famous farming areas as Brie and Champagne. Unlike Brandenburg, the Paris Basin is a prosperous growing region, but it too is undergoing major changes in farming and along its urban–rural fringe.

Rural change near Paris

Land use changes

Changes: It is at the edge of Paris where rural changes have been taking place most quickly. Beginning in the 1960s, **many motorways were built** radiating from Paris and linked with a ring-motorway (the *Peripherique*) that encircled

Paris. Now an outer ring-motorway is being built. In the 1970s, **a new airport was built** (Charles de Gaulle) north of Paris at Roissy to take pressure off Paris' Orly airport. Air transport has since become so popular that there are now plans to enlarge this airport.

Problems: But, building more motorways or airport terminals causes a lot of controversy because of the **environmental problems** they cause. (These are explained on page 70). For instance, environmental groups protested angrily when forests near Versailles were cut down to make way for a new motorway. Other people dislike new motorways and airport terminals because **they change the land uses around them**. These changes are shown in Figure 55.1. There is now **less farmland and less wildlife** because their habitats have been built on. The new motorways also attract so many new land users that **traffic increases enormously**. They become busier and busier and, despite being built to reduce congestion, are now actually the main congestion blackspots.

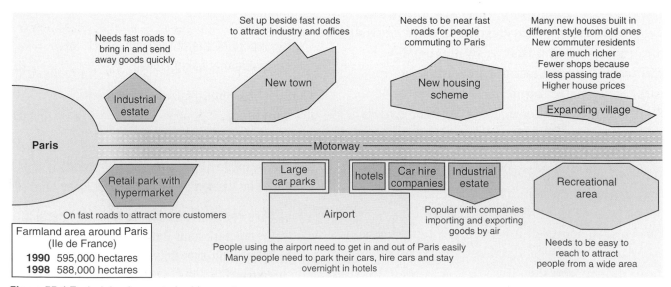

Figure 55.1 Typical developments beside a motorway

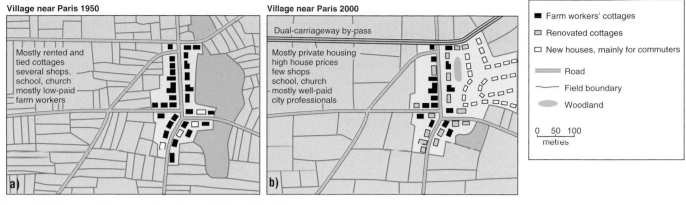

Figure 55.2 Typical village near Paris in **a)** 1950 and **b)** 2000

Policies: The authorities are now aware that building new motorways causes more problems than it solves. So they are **improving the railway network**. There are more railway lines and railway stations around Paris for people to use, as well as the Channel Tunnel rail link to Paris and the very fast TGV 'bullet trains' to major cities. To reduce urban sprawl, **developments have been concentrated on five New Towns and**, more recently, **in growth centres within Paris** (see Figure 54.1). These should reduce developments elsewhere in the urban–rural fringe, so that there will be less urban sprawl in other areas and fewer environmental problems.

Settlement changes

Changes: As transport improves, many villages in the urban–rural fringe are becoming **commuter settlements**, e.g. Meudon, Saint Denis. Living in the countryside is attractive to many people because it offers cheaper housing, better air quality, less crime and more peace and quiet.

Problems: But, as commuters move in, the village starts to grow and change. **Countryside is lost** as the village expands. The village often becomes less attractive as estates made up of almost identical houses are built, which are different in style from the old cottages. The newcomers sometimes do not fit in with the local villagers. They have different interests and occupations and are wealthier. If a by-pass is built, shops may close because there is no longer any passing trade and the commuters mostly shop

in the city. In these ways, **the character and community spirit in the village disappears**.

Policies: Local authorities can **establish green belts around villages** to prevent them from growing any further. They can also prevent the character of the village changing by **conserving old buildings**. In addition, **by making Paris a more attractive area in which to live**, this should mean that fewer people will want to move to the countryside.

QUESTIONS

1 What new transport links have been built recently near Paris? **(1)**

2 Name four land uses often found beside these new transport links **(4)**

3 Describe the problems caused by these new forms of transport **(2)**

4 Describe how the railway network around Paris has been improved **(3)**

5 Explain how a better railway network reduces problems in this region **(2)**

6 Describe two problems that arise when a village becomes a commuter settlement **(4)**

7 Describe ways in which these problems can be reduced

Look at Figure 55.2.
Describe the changes in the village between 1950 and 2000 **(5)**

56 Rural change in the Paris Basin, France (2)

Changes in the countryside

Farming changes

Changes: The Paris Basin is the most fertile and prosperous farming area in France. Many types of farming are profitable here, including dairying in Brie and growing vines in Champagne. But, with its warm, sunny climate and loess soils, cereal growing is most important. It has become particularly popular since the European Union set up its Common Agricultural Policy. This has led to many changes in the farming landscape of the Paris Basin (see Figure 56.1).

Problems: The changes brought about by the EU's Common Agricultural Policy have, on the whole, benefited farmers. Cereal growing, especially, is more profitable than ever before. But **many of these changes have been harmful to the environment**. Using more chemicals can kill the soil organisms, which help to keep the soil fertile. This means that more and

more chemicals have to be applied to maintain soil fertility. Some of the chemicals can be washed into rivers, killing wildlife there, or even into water supplies for villages and towns.

Fewer hedges and trees mean that winds are stronger. After harvest, when the soil is bare, these winds can pick up and blow away soil. Over many years, the fertile topsoil is eroded, the soil becomes thinner and crops do not grow as well.

There has been a great reduction in wildlife in this area as their habitats, such as trees, wetlands and hedges, are all disappearing.

Policies: Because of this environmental damage, the European Union has recently brought in **new environmentally-friendly policies**.

1. Farmers are now encouraged to practice organic farming. Grants and advice are available to help farmers grow crops without using artificial chemicals.
2. By paying farmers to take out of production some of their arable land (*set-aside* policy), this

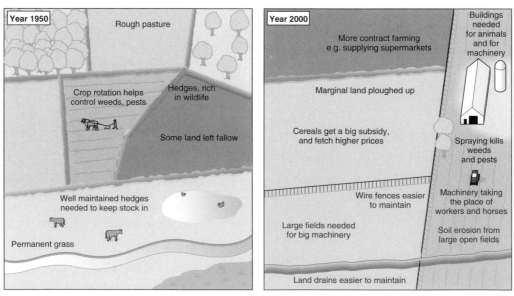

Figure 56.1 Land-use changes in the Paris Basin (1950–2000)

should increase wildlife and decrease the amount of chemicals used.

3. Information is also given on how to prevent and cure soil erosion. Farmers are advised to plant shelter belts and to practice more mixed farming. If the farmers rear some animals, the extra fields of grass will mean fewer bare fields and less soil erosion. By manuring and by crop rotation, the farmer can also keep his fields fertile without using large amounts of artificial chemicals.

land use	1990 (million ha)	1998 (million ha)
cereals	4.92	4.12
woodland	3.48	3.66
grassland	2.95	3.32

Figure 56.2 Land-use changes in the Paris Basin (1990–1998)

Housing changes

Changes: Well beyond the urban–rural fringe, up to 100 kilometres from Paris, villages with a fast rail connection are becoming **commuter settlements** (see page 115).

In less accessible parts of the Paris Basin, **houses are being bought as** *second homes* **by people living in Paris**. About 20% of French households now have a second home and the Paris Basin is a very popular area, e.g. Seine et Marne (near Fontainebleau), Yvelines (the Chevreuse valley). In some areas there are more second homes than farms and, in some villages, there are more houses owned as second homes than by permanent residents.

Problems: As second homes take over in a village, **local services close down** because the second home owners only stay there at weekends and holidays and so do not use the services enough. Shops, even schools, close down and the bus service is reduced.

Although the second homes remain empty for much of the year, local people who wish to live in the village cannot do so because they cannot afford the **much higher house prices**.

Policies: In response to these difficulties, the French government is trying to **improve living standards in the countryside**. The government has increased agricultural training facilities and agricultural schools so that farming is more efficient. Farmers can also now obtain loans more easily and there are grants given to young farmers who take over a farm from a retired farmer.

Many farmers in popular holiday areas, such as Champagne, cater for tourists. They have **changed old farmworker cottages into holiday homes** (or *gites*). This brings the farmer extra income.

There has also been an **increase in industries setting up in the countryside**. This is because companies have to pay less tax if they move out of Paris. By these policies, it is hoped that fewer people will wish to leave the countryside, so there will be fewer empty houses to be bought up by city people and the village community will be stronger and more prosperous.

QUESTIONS

1. Describe how farming in the Paris Basin has changed since the 1950s (4)

2. Explain how growing cereals causes environmental problems (4)

3. Describe two new policies in the Paris Basin and explain how each one improves the environment (4)

4. What problems do second homes cause in the Paris Basin? (4)

5. Explain how the French government is improving living standards in the countryside (6)

6*. a) **Look at Figure 56.1. List seven changes that have taken place in farming in the Paris basin since 1950**

   b) **Look at Figure 56.2. List three changes that took place in the 1990s** (5)

57 Measuring development (1)

Introduction

This unit studies important geographical issues on a world scale, and none is more important than the huge differences in health and wealth from country to country.

Most countries are trying to improve the conditions in which their people live. **Any improvement that is made in the standard of living of the people is called** *development*.

Some countries have developed more than others. Their people enjoy a high standard of living. These are the *Developed Countries*. Those that have not developed as much are called *Developing Countries* or *Less Developed Countries*. Their people have a much lower standard of living.

Measuring development

It is very difficult to work out one person's standard of living. To try and measure precisely the standard of living of all the people in a country is impossible. The best that can be done is to select a few indicators of development and measure these e.g. people's income, life expectancy, education, food intake. Three types of development indicators are studied here.

Figure 57.1 This large house and expensive car demonstrates an affluent lifestyle and a high standard of living, in economic terms

Figure 57.2 In this shanty town in Natal, South Africa, conditions are poor, and the standard of living is low

Economic indicators of development

These have been the most commonly used indicators. **They measure the wealth and industrialization of a country**. Examples include:

Gross domestic product (GDP) per person

The *gross domestic product (GDP)* is the value of all the goods produced and services provided in a country in one year. This is divided by the number of people living in the country to indicate the wealth of the average person.

Gross national product (GNP) per person

The *gross national product* is similar to the GDP, but it also includes services earned abroad.

Energy used per person

The amount of energy (coal, oil, gas, etc.) that is used in a country can also indicate economic development. Countries with a lot of industries producing much wealth will also use a lot of energy. People with a high standard of living will use a lot of petrol in their motor cars and much electricity in their homes.

People employed in agriculture

A country with a high proportion of its people in agriculture will have little industry to produce wealth. In addition, its farms are likely to be small and unprofitable. So, a high percentage of people in agriculture is a good indicator of a less developed country, and vice versa.

Problems with economic indicators

- Although a country may produce a lot of wealth, it may not be spread out amongst all of its people. A small number may be extremely wealthy while the vast majority remain poor.
- The amount of wealth does not give enough information on people's quality of life, e.g. how healthy they are, how well educated.
- The amount of income and wealth does not even show how well-off the people are. This needs to be compared with prices to find out what people can buy with that amount of money.

Social indicators of development

Social indicators show how a country uses its wealth to improve the quality of life of its people. Those that measure health include:

- population per doctor
- infant mortality (the number of children who die before they are one year old)
- life expectancy

Those that measure diet include:
- calories per person per day
- protein per person per day

Those that measure education include:
- percentage of children attending secondary school
- adult literacy

Problems with social indicators

- They also use averages, so they do not tell us the differences within a country. For example, the average number of calories per person might be 2500 per day, but half of the people might only receive 2000 calories and be severely undernourished, while the other half have 3000 calories and be well-fed.
- One indicator on its own does not give enough information on quality of life. Although people may be well-fed, we do not know how healthy or well educated they are.

QUESTIONS

1. What is meant by 'development'? (2)

2. Describe two economic indicators of development (4)

3. Name four social indicators of development (2)

4. What is the difference between an economic and a social indicator of development? (2)

5. Most indicators give average figures. Describe the problems with using averages (3)

Look at the table below then answer these questions
   a) **What is the best way of measuring the standard of living of people in France and the UK? Give reasons for your answer**
   b) **Which country is more developed—Britain or France? Give reasons for your answer (5)**

Indicators of development	France	UK
average income ($)	24,990	18,700
life expectancy (years)	79	77
population per doctor	333	300
calories eaten per person each day	3633	3317
number of cars per 1000 people	430	373
% people working in agriculture	6	2

58   Measuring development (2)

Comparing social and economic indicators

Generally, countries that score highly on economic indicators also do well according to social indicators. This is because they can use their wealth to provide proper schooling, hospitals, food and decent housing. Countries with little wealth just cannot afford to provide all of these social services for their people.

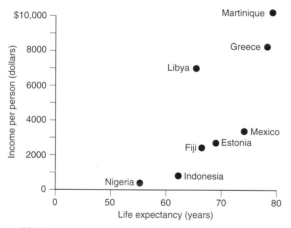

Figure 58.1 Income per person and life expectancy in selected countries (1995)

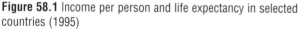

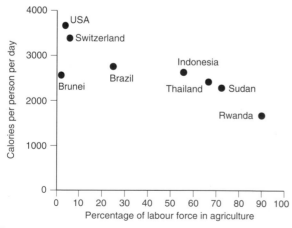

Figure 58.2 Calories eaten per day and people working in agriculture in selected countries (1992)

Some countries, however, appear more developed according to social indicators whereas others appear more developed according to economic indicators. This is shown in Figure 58.1 and 58.2. For example, Figure 58.1 shows that generally as a country's income per person increases, so does the life expectancy of its people. But, some exceptions can be spotted. People in Libya have a lower life expectancy than people in Mexico, but their average income is twice as high.

country	income per person ($)	adult literacy (%)
Chad	180	30
India	340	50
Algeria	1600	57
Fiji	2440	90
Lebanon	2660	91
St Lucia	3370	93
Hungary	4020	99
Oman	4820	35

Figure 58.3

country	GNP per person ($)	population per doctor
Guyana	590	6000
Morocco	1110	4500
Tonga	1630	2000
Tunisia	1820	1852
Turkey	2780	1176
Mexico	3320	621
Uruguay	5170	500
Georgia	440	182

Figure 58.4

Combined indicators of development

Because different indicators give different results,

it is more reliable to use several indicators. Often, **a range of social and economic indicators are used**. For example, to compare the development of the two most populous countries in the world, five indicators have been used (Figure 58.5). According to four of these, China is more developed than India. Alternatively, **a range of indicators can be used to produce a single combined index**. Two examples are:

● **The Physical Quality of Life Index (PQLI)**

This combines life expectancy, infant mortality and adult literacy to produce an index from 0–100. The higher the PQLI the higher the quality of life of the country.

● **The Human Development Index (HDI)**

This combines life expectancy, adult literacy, GNP/person, cost of living and school enrolment to produce an index from 0 to 1.

Indicator Of Development	China	India
GNP per person ($)	620	340
energy used per person (tonnes)	0.35	0.35
life expectancy (years)	71	61
calories per person per day	2727	2395
adult literacy (%)	70	50

Figure 58.5

Highest PQLI	Highest HDI
1. Sweden	Canada
2. USA	Norway
3. Norway	USA
4. Canada	Japan
5. United Kingdom	Belgium
6. Australia	Sweden
7. Japan	Australia
8. Denmark	Netherlands
9. Iceland	Iceland
10. Switzerland	United Kingdom

Figure 58.6 The World's most developed countries

QUESTIONS

1 Describe the advantages of combined indicators of development over economic or social indicators (2)

2 Explain how **(a)** the Physical Quality Of Life Index, and **(b)** the Human Development Index are worked out (4)

3 From the information in Figure 58.6, do you think the PQLI and the HDI give similar results? (3)

Techniques Questions

4 Look at Figure 58.1
   a) What is the life expectancy and income per person in **(i)** Greece **(ii)** Nigeria? (2)

5 Look at Figure 58.2
   a) What is the percentage of people in agriculture and the number of calories eaten per person in **(i)** Switzerland, **(ii)** Brazil, **(iii)** Sudan? (3)
   b) Which country has a higher percentage of people in agriculture—Rwanda or USA? (1)
   c) Which country has a higher number of calories eaten per person—Rwanda or USA? (1)
   d) What is the relationship between the percentage of people in agriculture in a country and the number of calories eaten per person? (2)
   e) Which country is an exception to this relationship, and how is it an exception? (2)

6 a) Draw a scattergraph to show the relationship between income per person and literacy rate in the eight countries listed in Figure 58.3 (3)
   b) Describe the relationship shown by the scattergraph (2)

7 a) Draw a scattergraph to show the relationship between the GNP per capita and the population per doctor for the eight countries listed in Figure 58.4 (3)
   b) Describe the relationship shown by the scattergraph (2)

59 Reasons for differences in development levels (1)

Global variations in development

By using one or more development indicator, the world can be divided into developed and developing countries. These are shown in Figure 59.1. The developed countries (or *the North*) are fewer in number, nearly all are in the northern hemisphere, and most are in temperate latitudes. The developing countries (or *the South*) have 75%

of the world's people. They are found in both the northern and southern hemisphere and they include all the countries within the tropics.

Figure 59.1 also shows that there are big differences in living standards within the developing world. For example, the average income in Argentina is 80 times that in Ethiopia and people can expect to live 30 years longer.

There are many reasons for the huge variations in standards of living around the world. The factors involved can be divided into physical and human.

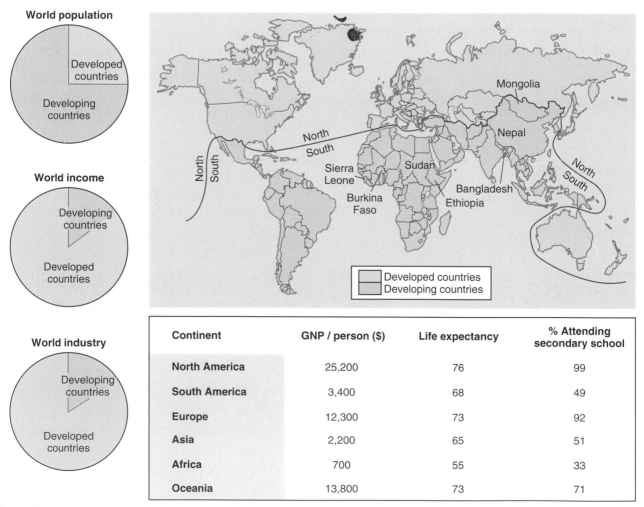

Continent	GNP / person ($)	Life expectancy	% Attending secondary school
North America	25,200	76	99
South America	3,400	68	49
Europe	12,300	73	92
Asia	2,200	65	51
Africa	700	55	33
Oceania	13,800	73	71

Figure 59.1

Physical factors

Factor	Problem	Why this is a problem	Example
climate	very cold	• difficult to build roads and railways • remote and unlikely to attract much industry • also too cold to farm • expensive to live because high heating bills, food is expensive, • houses difficult to build because of permafrost	Mongolia
	very dry	• barely enough rain to grow crops • always a risk of crop failure and famine • remote and unlikely to attract industry • soil made poorer by wind erosion	Ethiopia
relief	very steep	• also difficult to build roads and railways, so remote and unlikely to attract much industry • poor farming because of steep land, inability to use machinery and thin soils	Nepal
resources	lack of minerals	• no valuable minerals (e.g. diamonds, gold) to sell to other countries • no fuels (e.g. coal, oil) to encourage industry to set up	Sudan
environment	unattractive scenery	• not attractive to summer tourists (e.g. no sandy beaches, hot, sunny climate) or winter tourists (e.g. no cold, snowy, steep slopes)	Burkina Faso
	much disease	• a country is unable to develop if many of its people suffer from disease and are unable to work property (see Chapter 61)	Sierra Leone
natural disasters	floods, drought, earthquakes, volcanic eruptions, hurricanes	• areas prone to natural disasters have harvests ruined, factories and homes destroyed, roads and railways unusable • costs millions of pounds and may cause famine and unemployment • may take years for the areas to recover	Bangladesh (see Chapter 85)

QUESTIONS

1. Describe the world distribution of developing countries (2)

2. Compare life expectancies *within* the developing world (3)

3. Explain how climate can affect the development of a country (4)

4. Describe the importance of relief in explaining the development level of a country (2)

5. Explain how resources and environment help to explain differences in standards of living around the world (4)

Choosing from:

very cold, very dry, mountainous, prone to severe earthquakes, suffers from frequent eruptions, affected by frequent hurricanes, which physical problem do you think a country finds it easiest and most difficult to overcome? Give reasons for your answer (5)

60 Reasons for differences in development levels (2)

Human factors

Some countries find it difficult to develop because of their physical environment, but there are many states that have overcome the problems of a harsh environment and enjoy a high standard of living. Such countries include Japan, Finland, Switzerland, Canada and Australia. There must, therefore, be other factors—human factors—that help to explain differences in development levels around the world. Some of the most important of these are described below.

Population growth

	developing world	developed world
average birth-rate (%)	27	12
average death-rate (%)	9	10
average natural increase (%)	18	2

Figure 60.1

As Figure 60.1 shows, **population is rising nine times faster in developing countries than in the developed world**. This gives poorer countries two sets of problems.

In the countryside, **farms become smaller**, as there are more people needing land. So the farmers produce less food for their families to eat and have an increased risk of going hungry.

In the cities, **the city authorities cannot provide enough houses, schools, hospital beds and jobs for the increasing population**. So many people live in makeshift houses (*shanty towns*), are underemployed and have little chance of getting to a hospital if they are ill. In Niger in west Africa, for example, there are 54,000 people to every doctor.

Because the birth-rate is still high, there are many young children in developing countries. In Kenya, for instance, half of the people are 14 years old or younger. **This large number of children places an additional strain on the country**. The children do not produce wealth, but they need to be kept healthy, well-fed, educated and properly clothed.

Industrialization

	developing world	developed world
% people working in:		
agriculture	63	5
manufacturing	12	30
services	25	65
% of world's industry	15	85

Figure 60.2

As Figure 60.2 shows, **there are far fewer factories and offices in developing than developed countries**. Factories and offices produce profits that increase a country's wealth. They also employ many people, providing them with a regular wage. Without industry, a country finds it very difficult to develop.

In addition, although there is little industry, the population in developing world cities is rising rapidly. This means that more and more people are unemployed or underemployed and have only a low standard of living.

	developed countries	developing countries
imports	manufactured and primary goods	expensive, manufactured goods
exports	expensive manufactured goods	cheap, primary goods
trade balance	trade surplus	trade deficit
debts	lend money to poorer nations	borrow money at high interest rates

Figure 60.3

Factories and offices are less likely to set up in developing countries because there are few people there who are rich enough to buy their products. So the goods have to be transported great distances to be sold, which increases costs. The roads and railways are also poorer and there are fewer banks from which to borrow money. With fewer secondary schools and universities, there are not many people with the necessary skills (e.g. in information technology) to work in a modern office. Although some industries are found in poorer countries, they are often foreign-owned (*multi-national companies*), so the profits do not stay in that country to increase its wealth.

Trade

As Figure 60.3 shows, with few factories, **most developing countries have only primary goods to export** (such as crops and minerals). **Their prices are generally low** and also fluctuate greatly. For example, in the mid–1990s, the world cocoa price was 60% less than ten years before. **Developing countries need to import manufactured goods, but they are expensive** and generally rise in price. So, the money they receive from their exports does not usually pay for their imports. This means **they cannot afford to provide enough services** (e.g. hospital equipment, school books, agricultural machinery) to enable people to enjoy a higher standard of living. It also means that, over the years, they have borrowed large amounts of money from developed countries and now **spend much of their income just in repaying interest on these debts**. This is money that otherwise would have been spent on improving people's standard of living.

Developing countries even find it difficult to export the few goods that they do produce. This is because other countries put up *trade barriers* to protect their own industries. So, developing countries may find that they are only allowed to export a limited number of goods to countries such as the USA (a *quota*) or find that a tax or *tariff* is put on their goods so that their price is too high for people to buy.

QUESTIONS

1 a) Describe the differences in birth-rates and death-rates in developed and developing countries **(2)**

b) How do population problems explain the low level of development in Developing countries? **(4)**

2 a) Give four reasons why fewer factories set up in developing countries **(4)**

b) Explain why countries with many factories and offices can enjoy a higher standard of living **(2)**

3 a) Describe the differences in imports and exports in developed and developing countries **(4)**

b) Explain why most developing countries are in debt **(2)**

c) How do trade problems explain the low level of development in poorer countries? **(6)**

4* **For a typical developing country, which of the following would do most to improve the standard of living of its people?**
- **halve its birth-rate**
- **double its exports**
- **double its factories**

Give reasons for your answer **(5)**

61   Health problems in the developing world

We have already found out that there are many ways of measuring or indicating a country's level of development. The quality of people's health is one of these indicators. If many of a country's people are suffering ill-health, this indicates a low level of development. But it is also a major *cause* of a low level of development. Differences in health are one of the most important reasons why some countries have developed more than others. In this chapter we shall look at the problems of ill-health faced by people in the developing world.

Causes of ill-health

Diseases can be divided into those that are *infectious* (where one person infects another) and *non-infectious* (which cannot be 'caught' from someone else). In the developing world, infectious diseases are far more common and account for most people's cause of death.

Figure 61.1 shows the most common ones in each category.

Figure 61.2 The blackfly bites people and spreads river blindness

infectious diseases	diseases spread by water, e.g. cholera, diarrhoea, typhoid, snail fever	These are the most commonly-occurring diseases. They are spread in two main ways: 1. People drink polluted water. The water has usually become polluted by human sewage. When people drink polluted water, they also swallow the invisible bacteria, which cause a variety of diseases. 2. Tiny worms that live in water burrow into people's skin when they are washing, fishing or playing in the water. These worms then grow and multiply inside the human victim and cause serious illnesses.
	diseases spread by flies, e.g. malaria, river blindness, sleeping sickness	Areas that are hot and wet are ideal environments for many flies and mosquitoes. They spread disease by biting people and, as they do so, bacteria or tiny worms, pass from their saliva into the person's bloodstream and they become ill.
	Others	e.g. leprosy, AIDS, measles, TB
non-infectious diseases	diet-deficiency diseases, e.g. rickets, under-nutrition, kwashiorkor, scurvy	These diseases are caused by people not eating enough food or not eating a balanced diet of carbohydrate, protein, vitamins and minerals. This is mostly caused by (a) poor harvests, (b) not owning animals to provide protein, (c) being too poor to eat a proper diet.
	Others	e.g. heart diseases, cancers, accidents, brain diseases

Figure 61.1 Diseases of the developing world

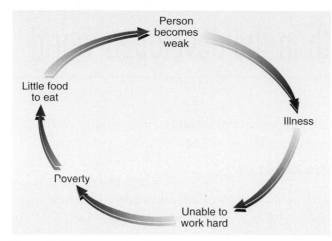

Figure 61.3 Vicious cycle of disease

Effects of ill-health

The diseases common in the developing world affect people in a variety of ways.

1. **Some are killer diseases**, e.g. AIDS, malaria.
2. Others are much less likely to kill, but **they make people very weak and lethargic**, e.g. snail fever, kwashiorkor. If people are very weak, they quickly become trapped in the vicious cycle of disease (see Figure 61.3).
3. Some diseases **leave their victims with a permanent handicap or injury**, e.g. river blindness, rickets. Not only does this make it extremely difficult for the people affected to work, it also requires someone in the community to look after them.

In developing countries most people suffer from at least one disease. This seriously reduces their quality of life but it also **seriously reduces the economic development of the whole area** because (a) the people who are unwell cannot work and produce wealth, and (b) they also need other people to look after them—people who otherwise would be working themselves.

Methods of control

Most of the diseases in the developing world can be prevented or cured. There just isn't enough money to do so. The money is needed to:

1. **Produce more food**—this reduces diet-deficiency diseases and makes people stronger and better able to fight off other diseases,
2. **Improve health facilities**—so that everyone is within reach of and has access to some form of health care, even people living in the countryside
3. **Provide clean water**—by improving water supplies and sewage disposal
4. **Provide health education**—so people know what causes diseases and simple ways in which they can be prevented

QUESTIONS

1 Name the two main ways in which infectious diseases are spread **(1)**

2 Explain how polluted water spreads disease **(2)**

3 Explain how flies and mosquitoes spread disease **(2)**

4 What is meant by diet-deficiency diseases? **(2)**

5 Give two examples of diet-deficiency diseases **(1)**

6 Describe in words the vicious cycle of disease **(3)**

7 Explain how disease slows down the development of a region or country **(4)**

8 Describe three ways in which diseases of the developing world can be controlled **(6)**

Four methods of controlling disease are listed above.
Put the four methods into order according to:
a) **how costly they would be**
b) **how good they would be in reducing disease**
Which method is the best for a poor country?
Give reasons for your answer **(5)**

62 Problems of ill-health in the developed world

In the developing world, infectious diseases are more common chiefly because of the lack of money to prevent and cure them. In developed countries, a hundred years ago, the situation was very similar. Infectious diseases, such as whooping cough, diphtheria, scarlet fever, tuberculosis, typhoid and measles were rife. Since then, however, we have learned how to prevent and cure them and they now account for relatively few deaths. Instead, it is the non-infectious diseases that are now more serious in the developed world (see Figure 62.1) These are much more difficult to prevent and cure and this is mostly because we do not fully understand their causes.

Causes of ill-health

Four factors are particularly important in explaining the diseases of the developed world.

Pollution

Air pollution from vehicles, factories and power stations is more serious in developed countries.

This causes many diseases, such as lung diseases and some forms of cancer.

Social habits

Habits such as smoking and drug and alcohol abuse increase our changes of dying from a variety of diseases, such as lung cancer and liver diseases. These habits are expensive and so are more common in richer countries

Diet

Unlike the developing world where people suffer from a lack of food, in richer countries people suffer ill-health from eating too much, especially fatty foods. This puts us at risk of heart diseases and cancers.

Stress

The faster, more hectic pace of life in cities of the developed world affect our health. Stress is linked to heart diseases, brain diseases, even accidents and suicides.

Effects of ill-health

Although people in the developed world are much healthier than in poorer countries, we still need to **spend a lot of money on health care**. About 6% of all the government's money is spent on health, e.g. on doctors, hospitals, equipment and drugs. And as the equipment and drugs become more advanced, they also become more expensive.

Absenteeism from work because of illness can also **slow down production in factories and offices**. This happens especially when there are flu epidemics in winter and a large number of employees can be away from work at the same time. For example, a flu epidemic in Scotland in

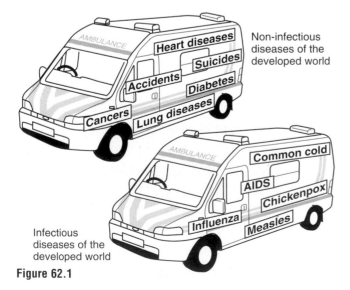

Non-infectious diseases of the developed world

Infectious diseases of the developed world

Figure 62.1

January 2000 cost businesses millions of pounds as one in five workers were too ill to go to work.

Overall, however, the richer countries have been very successful in reducing ill-health so that people can now expect to live much longer. In 1900, people in Britain could expect to live until they were 52 years old. Now this figure is 79 years, although a little less for males than females. This gives us an additional problem of **an ageing population**, the effects of which were discussed in Chapter 46).

Controlling disease

There are three main ways in which richer countries have reduced disease.

Improved environmental conditions

People in developed countries live in a cleaner, healthier environment. Our water is purified before it reaches our taps. Sewage is taken away by pipes and treated before being emptied into rivers and seas. Rubbish is collected regularly and disposed of properly. Under these sanitary conditions, infectious diseases are much less likely to spread.

Better health facilities

As already mentioned, richer countries can afford to spend a lot of money on medical care. A full range of equipment and help is available, from by-pass operations, kidney transplants, fully-trained midwifes, to physiotherapists and ante-natal care. Children are routinely inoculated against diphtheria, tetanus and polio when they are a few months old; against measles, mumps and rubella when one year old, and against tuberculosis when about 13 years old. To ensure that we are not affected by the vicious cycle of disease, sickness benefit is paid to people who are unable to work through illness.

Health education

People in the developed world are much more aware of the causes of disease and how they can

Figure 62.2 This is a still from a TV advertisement to combat drink driving – a big problem in the developed world

be prevented. We know the importance of a healthy diet, regular exercise and safe sex. It is much easier to inform people of health matters in richer countries. Countless radio programmes and TV chat shows discuss topical health issues. The health services regularly run campaigns and they get their message across through advertisements in newspapers, on television (see Figure 62.2), in schools and on roadside billboards.

QUESTIONS

1. Name four non-infectious diseases common in developed countries **(2)**

2. Explain how a poor diet in richer countries can cause disease **(3)**

3. Describe the type of social habits that can lead to ill-health **(2)**

4. Explain why developed countries need to spend an increasing amount of money on health care **(3)**

5. Explain why infectious diseases spread by water are unlikely to affect people in the developed world **(4)**

6. Describe the health care given to a mother and her baby before and after the child is born **(4)**

7. **Give examples of health messages you can remember. How were they advertised (e.g. on TV, on posters) and how good were they? (5)**

63 Diseases of the developing world—malaria (1)

We have already found out that infectious diseases are more common in the developing world and non-infectious diseases in the developed world. To show the effects of these diseases, the most common ones affecting both developing and developed countries are now studied in more detail. The first disease affects 400 million people and kills two million every year, half of them being children. This disease is called **malaria**.

Cause and method of transmission

Malaria is caused by a tiny parasite that finds its way into a person's bloodstream. After a few days, the infected person suffers headaches and stomach pains, followed by fevers of high temperature and shivering fits. The fevers can occur many times, frequently resulting in the death of the victim. Malaria is a particularly big killer of children, who have not had time to build up any immunity from the disease. If malaria does not kill the victim, it can cause kidney failure. It leaves the patient weak and anaemic and prone to other diseases. The person's life expectancy is reduced considerably.

In areas where malaria occurs, many of the people will have the disease. As a result, the amount of wealth (from farms and factories) that the area produces is seriously reduced while, at the same time, a lot of time and money has to be spent on caring for all the victims. In the Philippines, for example, when malaria was rife in the 1940s, absenteeism from work was 35%. In regions where malaria is particularly bad, people have been forced to move away, leaving behind fertile farmland, e.g. in northern Sri Lanka.

The tiny parasite that causes malaria enters a person's bloodstream when he/she is bitten by a mosquito. Not all mosquitoes carry the disease. Only **the female anopheles mosquito spreads malaria**. It bites an infected person and sucks blood containing the parasite into its stomach, where the parasites multiply. The mosquito then bites someone else and the parasite enters the new victim on the saliva of the mosquito.

QUESTIONS

1. What causes malaria? **(1)**
2. What spreads malaria? **(1)**
3. Describe how malaria is spread **(3)**
4. Describe how malaria affects its victims
5. Explain how malaria slows down the economic development of a region **(6)**
6. Look at Figure 63.2. Describe the distribution of malaria in the 1940s **(2)**
7. Look at Figures 63.3 and 63.4 Compare the distribution of malaria in the 1970s and 1990s **(4)**

    Look at the three maps on page 131. Take each continent in turn and state how the area affected by malaria has changed between the 1940s, 1970s and 1990s **(5)**

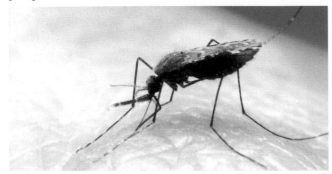

Figure 63.1 A mosquito taking a blood meal

Distribution of malaria

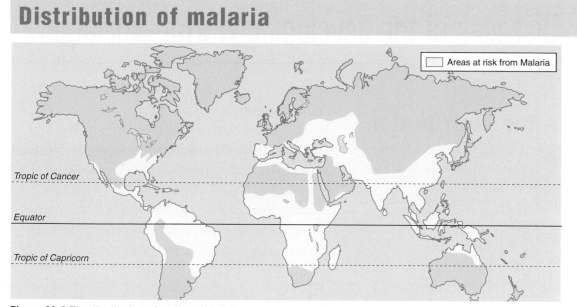

Figure 63.2 The distribution of malaria (1940s)

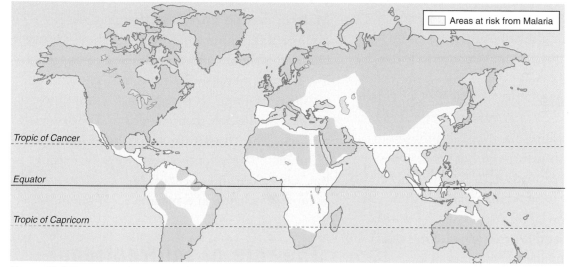

Figure 63.3 The distribution of malaria (1970s)

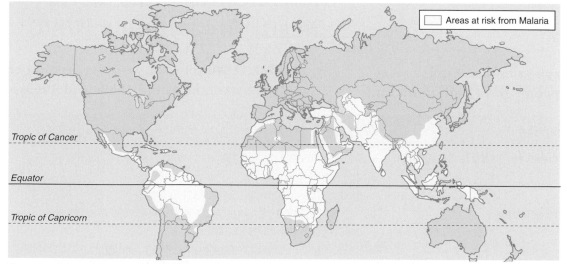

Figure 63.4 The distribution of malaria (1990s)

Diseases of the developing world—malaria (2)

Factors in the distribution of malaria

Malaria has occurred in most areas of the world at some time. As Figure 63.2 shows, in the 1940s even people in the United States and Europe were affected. Since then the disease has retreated (Figure 63.3) but in the 1980s and 1990s expanded once again (Figure 63.4). Now, malaria exists in 100 countries and nearly half the world's population live in malarious areas. The reasons behind its changing distribution are a combination of physical and human factors.

Physical factors

Malaria occurs where the anopheles mosquito lives. They live in warm and hot areas, where **temperatures are above 16°C**. They need **still water surfaces** as breeding areas, but these areas do not need to be large. As a result, all **warm, rainy areas** with still or slowly-moving water are suitable environments for the anopheles mosquitoes.

Human factors

People's activities also affect the distribution of malaria. **Where people have built dams and made irrigation channels**, they have created suitable breeding areas for the mosquito and so malaria increases. **People migrate much more now** and this makes it easier for the disease to spread. In some areas of the world, people have successfully used insecticides to kill off the mosquito. This explains why the areas affected by malaria decreased between the 1940s and the 1970s.

Controlling malaria

Malaria can be cured by drugs, such as quinine and chloroquine, which kill the parasite that causes the disease. But drugs are now much less effective because the parasites have built up resistance to them. It is also much cheaper to prevent the disease than to cure it. **To prevent malaria, the mosquito must be destroyed**. Until the 1970s scientists were very successful but, since then, the mosquito seems to be winning the battle and malaria is on the increase.

The role of international organizations: the World Health Organization (WHO)

- **The WHO launched a world campaign to eradicate malaria in the 1950s and 1960s.**

Methods of prevention	difficulties
1. spray insecticides (e.g. DDT, dieldrin)	• some chemicals pollute the environment, killing other life forms • they are very expensive • mosquitoes are becoming resistant to them
2. drain breeding grounds	• impossible to do thoroughly, as only small areas of water are needed, e.g. pot-holes in roads • water is also needed for drinking, washing, for crops and animals

It used drugs to cure people and insecticides to kill mosquitoes. This was very successful initially and malaria was reduced. But, the campaign finally failed because (a) it was very expensive for poor countries to buy insecticides, and (b) the mosquito was becoming resistant to the chemicals used. In India, for example, the WHO reduced malaria from 75 million cases in the 1950s to less than one million cases 10 years later. Unfortunately, two million people now have the disease.

- **The WHO employs scientists to find better ways of curing and preventing malaria**. They developed DDT and, when this became less effective, developed dieldrin. They have tried more ingenious methods, such as introducing sterile male mosquitoes and fish that eat mosquito larvae. However, only £40 million is spent researching ways of reducing malaria, compared with £600 million spent on AIDS research.

The role of aid agencies: the Red Cross

The Red Cross provides emergency medical help (e.g. drugs, equipment, nurses) when epidemics of malaria occur, but they too believe that prevention is better than cure. So they also provide **long-term medical help** to improve health conditions, especially in

countryside areas. **They take someone from each village and give them training in primary health care** (PHC). The person then returns and educates everyone else in the village on health matters.

Primary health care includes giving advice on how diseases such as malaria are spread and low-cost ways of reducing these diseases. These methods include (1) persuading people to use mosquito bednets, (2) covering water containers, (3) filling in puddles and (4) reducing visits to the river.

- In the Philippines, for example, the Red Cross have taken someone from each remote mountain village and given them a six-week training course to become their village health worker.
- In the Son Ha province of Vietnam, the Red Cross have set up a malaria control programme that is now managed by the local people themselves. So far, they have concentrated on health education. They have also distributed 16,000 mosquito nets to people in eight villages. These nets are soaked in insecticide and greatly reduce the risk of people being bitten by mosquitoes.

QUESTIONS

1. Describe the type of physical environment where malaria is likely to be found **(3)**

2. In what ways have people helped to spread the disease? **(2)**

3. Describe the methods by which the World Health Organization try to prevent malaria **(3)**

4. Explain why malaria is increasing worldwide **(3)**

5. Explain how the Red Cross try to prevent malaria **(5)**

6* **List all the ways, which have been mentioned on these two pages in which malaria can be cured or prevented** **(5)**

Figure 64.2 The Red Cross own mobile laboratories to provide medical help to those in more remote areas

65 Diseases of the developing world—cholera (1)

About 80% of all sickness and disease in developing countries is caused by drinking polluted water. Five million children die each year from diarrhoeal diseases caught in this way. One of the most feared of these water-borne, infectious diseases is cholera. In the 1800s it was a dreaded disease all over the world. Six major epidemics spread across the globe, including Britain, and wherever it struck many people died in a very short time. In 1854, one outbreak in part of London killed 500 people in one week. The disease then travelled to Oxford, where it did similar damage.

In the 20th Century, cholera retreated, but it still flares up occasionally. In 1947 20 000 people died when the disease struck Egypt.

Cause and method of transmission

Cholera is caused by a comma-shaped bacteria that lives in a person's intestine. It leads to uncontrollable vomiting and diarrhoea, so people lose many litres of water very quickly. This causes severe dehydration and the victims lapse into a coma. If untreated, over half the victims die within 24 hours.

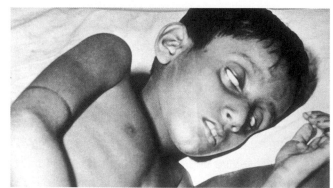

Figure 65.1 Cholera victim

Cholera is spread by polluted water. Sewage from people with the disease finds it way into a river or well. This sewage will contain millions of cholera bacteria. Other people then drink this water and swallow the invisible bacteria, and so become infected themselves. As the diarrhoea is so severe, it is highly likely that sewage from the new victim will also contaminate the area around and will get washed into water or even onto crops growing in fields, and so the disease spreads.

Distribution

In most areas of the world, cholera only happens in outbreaks or epidemics. The last major epidemic occurred in the 1960s. It spread quickly from south-east Asia to the rest of the world. By the 1970s there were even small outbreaks in Europe.

June 15, 1999
Cholera epidemic hits Cambodia
A deadly form of cholera has broken out in Ban Lung province which kills its victims within 6 hours of the first symptoms. 861 cases have been reported in the last month in 21 villages.

Figure 65.2

April 6, 1998
Cholera outbreak in Congo blamed on war
The recent cholera epidemic has been blamed on a breakdown in health standards following a five-month civil war. Tens of thousands of people were forced to leave their homes and flee to the port city of Pointe-Noire. Here they are living in overcrowded, unhealthy conditions where both water and food is contaminated.

Figure 65.3

Cholera often breaks out after major disasters, such as wars and earthquakes, when people no longer have access to safe water and are forced to drink polluted water.

Figure 65.4 shows that cholera is mostly found in Africa, but is also found in South America and in small areas of Asia. The great majority of the countries affected are within the tropics and all are in the developing world.

Physical factors

Physical factors do not play an important part in explaining the distribution of cholera. The bacteria can survive in most climates, except polar regions. **The disease is usually spread by water** and everyone, even those living in dry areas, must have access to this.

Human factors

It is the quality of the water that is crucial in the spread of cholera. **The water has to be polluted with cholera bacteria.** This happens **where there is poor sanitation** and human sewage goes untreated into rivers. People catch the disease when they drink polluted water directly from the river. This only happens in areas **where there is no clean, piped water available**. It spreads very quickly where there is **overcrowding** and many people have to use the same stretch of water as a toilet and a water supply. The disease is quite easy to control as long as medical help is nearby, so it is much more common in **areas where little medical help is available**.

QUESTIONS

1 What causes cholera? **(1)**

2 Explain how cholera is spread **(3)**

3 What are the effects of cholera? **(2)**

4 Using Figure 65.4, describe the areas affected by cholera in 1999 **(3)**

5 Name three human factors important in the spread of cholera and, for each, explain its importance **(6)**

Look at Figure 65.4. Put in order the 10 worst countries for cholera in the 1990s. Beside each one, write which continent it is in **(5)**

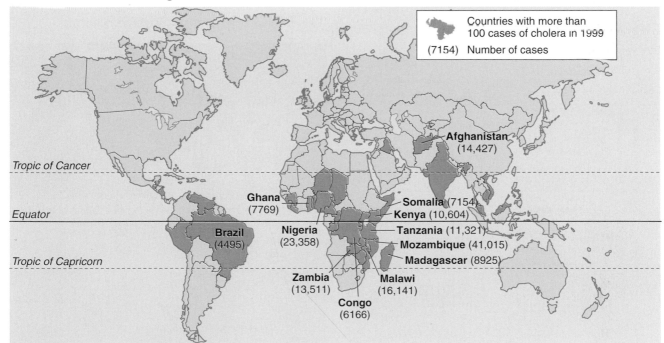

Countries with more than 100 cases of cholera in 1999
(7154) Number of cases

Afghanistan (14,427)

Tropic of Cancer

Ghana (7769)

Equator

Somalia (7154)
Kenya (10,604)

Brazil (4495)

Nigeria (23,358)

Tanzania (11,321)
Mozambique (41,015)
Madagascar (8925)

Tropic of Capricorn

Zambia (13,511)

Malawi (16,141)

Congo (6166)

Figure 65.4 World distribution of cholera

66 Diseases of the developing world—cholera (2)

Controlling cholera

To cure cholera, the victim must be given a lot of water, mixed with sugar and salt. If the victim's condition is already serious, then he must be given these fluids intravenously. This means that some medical facilities are needed nearby, such as needles, medicines and nurses.

To stop the disease from spreading, **the victim must be isolated** so that he cannot pollute local water supplies.

To prevent cholera, vaccines are available but they only offer partial protection. It is also too expensive for developing countries to vaccinate everyone and too late to vaccinate them once an epidemic breaks out. In the long term, **it is better to prevent cholera by improving sanitation**. This can be done by treating sewage or disposing of it safely (e.g. burying it). **Drinking water should also be cleaned** so that, even if bacteria are in the water, they will be killed off by the purifying process. As Figure 66.1 shows, more progress has been made with improving water supplies than improving sanitation in recent years.

The role of international organisations: the World Health Organization (WHO)

The WHO is part of the United Nations

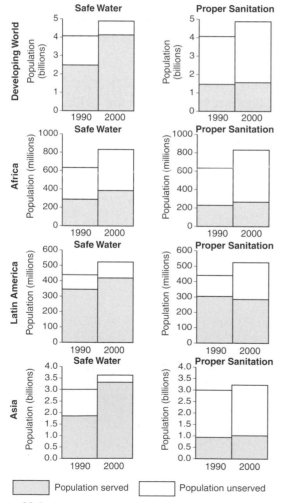

Figure 66.1

Figure 66.2 A demonstration to some Indian villagers of how to work the latrine, so as to improve sanitation

Figure 66.3 This waterpump in Nepal provides water that is clean and safe to drink

Organisation, from which it receives money for the work it carries out. There are two main ways in which it aims to reduce cholera.

1. **It trains local people to build and maintain their own improved water supply.** The WHO provides the money and technical assistance that is needed at the start of the scheme. For example, on the Indonesian island of Timor, people had to walk several kilometres to get their water, which often carried diseases. Cholera was commonplace. So, a well was dug in each village and a handpump installed (see Figure 66.3). Ten years later, cholera and other water-borne diseases have been almost wiped out. Income has trebled because people are much healthier and can spend more time working and less time fetching water. Someone in each village is trained to repair and maintain the handpumps.

2. **It improves sanitation in villages across the developing world.** It trains local villagers to build pit latrines (see Figure 66.2), which dispose of sewage safely.

The role of aid agencies: the Red Cross

When outbreaks of cholera occur the Red Cross try to reach the affected area quickly and **give victims the fluids they need to recover.** They need to arrive quickly as the victims can die within 24 hours of catching the disease.

After an earthquake in Turkey in 1999, the Red Cross and the Red Crescent organized camps for the homeless with water purification units, sanitation and electricity for around 20,000 people in less than one week. By acting so quickly, there were very few outbreaks of cholera.

To prevent cholera, **they take someone from each village and give them training in Primary Health Care** (see page 133). The person then returns to his/her village and educates everyone on simple and cheap ways of reducing cholera and other water-borne diseases, e.g. boiling water before use, disposing of sewage safely, advising against swimming or washing in infected waters, chlorinating water. They broadcast messages on radio and TV if possible, and put up banners and posters of information in public places.

In southern Sudan, the Red Cross is training local people in basic nursing skills including health education. It has also formed and trained teams of local workers to repair wells, handpumps and boreholes damaged during the recent war. This is improving water supplies and reducing the risk of people catching any water-borne disease, such as cholera.

QUESTIONS

1. Explain how cholera can be cured (2)

2. Explain how cholera can be prevented (4)

3. Look at Figure 66.1. Which continents in the developing world have the best and worst supply of
   a) safe water
   b) proper sanitation? (4)

4. Describe how the WHO improve village water supplies in developing countries (3)

5. Explain why pit latrines are a suitable way of improving sanitation in developing countries

6. Describe how the Red Cross try to prevent cholera in villages in the developing world

7* **Look at the graphs for the developing world in Figure 66.1. Did more or fewer people have access to**
   a) safe water
   b) proper sanitation in 2000 than in 1990?
   Answer the same question for Africa, Latin America and Asia (5)

67 Diseases of the developed world—heart disease (1)

Heart disease is the biggest cause of death in the developed world. It kills nearly half of all men and women. One in four men will have a heart attack before retirement age and most teenagers show signs of narrowing of blood vessels, which is the start of heart disease. But, unlike major diseases in the developing world, heart disease is non-infectious. One person cannot infect another. The causes are more complicated.

Causes

There are several heart (*cardiovascular*) diseases, such as strokes, angina and heart attacks. Some affect the arteries (which carry blood from the heart to the rest of the body). Others affect the heart itself. Many factors contribute to these heart diseases.

1. **Fatty diet: Too many fatty foods increases** *cholesterol*, which is a type of fat found in the blood. This narrows the arteries, increasing the chance of heart disease. Fatty foods also lead to people becoming obese or overweight, which puts an extra strain on the heart.

2. **Lack of exercise: Lack of exercise raises blood pressure and cholesterol** levels and can also cause the person to become overweight.

3. **Smoking: Nicotine increases your heart rate and blood pressure**, so more oxygen is needed for the heart to work properly. But smokers lose oxygen while smoking, putting the heart under extra strain. A packet of cigarettes a day doubles your chances of having a heart attack and makes you five times as likely to have a stroke.

4. **Stress: Stress increases a person's blood pressure** and so puts extra pressure on their

heart. Also, people under stress often indulge in 'comfort eating', e.g. chocolate bars or greasy chips, which can cause heart disease themselves.

5. **Inheritance: People can inherit a high blood pressure and high cholesterol** levels from their parents.

QUESTIONS

1 Describe the effects of smoking on your heart **(3)**

2 Explain how eating fatty foods increases your chances of heart disease **(4)**

3 Name the other main factors responsible for heart disease **(1)**

4 In what way does stress increase your chances of heart disease? **(3)**

5 Look at pictograph, Figure 67.1.
What percentage of people in the UK die from
a) heart disease
b) cancer? **(1)**

6 Look at pictograph, Figure 67.2.
What is the death-rate from heart disease in:
a) Latvia
b) Russia? **(1)**

7 Draw a pictograph to show the information in Figure 67.3.

Look at Figures 67.4 and 67.5.
a) Name the four areas in Scotland with the highest death-rate from heart disease
b) Describe their location **(4)**

a) Name the two areas in Scotland with the lowest death-rate from heart disease
b) Describe their location **(4)**

Cause of Death	Each symbol represents 5% of the total number of deaths in the UK
Heart diseases	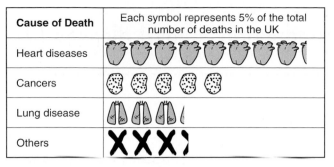
Cancers	
Lung disease	
Others	

Figure 67.1 Causes of death in th UK

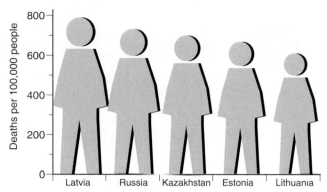

Figure 67.2 Countries with the highest death rate from heart disease (1995)

country	death-rate per 100,000 people
1. **Ireland**	367
2. **Finland**	340
3. **UK**	314
4. **Sweden**	235
5. **Germany**	231
6. **Greece**	175
7. **Portugal**	125
8. **France**	92

Figure 67.3 Countries with the highest death-rate from heart disease in Europe (1995)

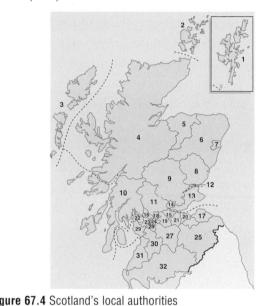

Figure 67.4 Scotland's local authorities

number on figure 67.4	Scottish local authority	Death Rate per 100 000 men*	Death Rate per 100 000 women*
1	Shetland	193	63
2	Orkney	166	70
3	Western Isles	234	55
4	Highland	187	59
5	Moray	165	78
6	Aberdeenshire	157	57
7	City of Aberdeen	173	67
8	Angus	162	62
9	Perthshire & Kinross	155	62
10	Argyll & Bute	194	72
11	Stirling	182	74
12	Dundee City	182	74
13	Fife	194	78
14	Clackmannanshire	219	78
15	Falkirk	210	89
16	West Dunbartonshire	229	88
17	East Lothian	162	60
18	East Dunbartonshire	133	58
19	North Lanarkshire	249	106
20	City of Edinburgh	170	57
21	West Lothian	215	76
22	Inverclyde	254	105
23	Renfrewshire	216	75
24	City of Glasgow	260	99
25	Scottish Borders	141	58
26	Midlothian	180	77
27	South Lanarkshire	220	90
28	East Renfrewshire	136	55
29	North Ayrshire	219	84
30	East Ayrshire	222	86
31	South Ayrshire	193	81
32	Dumfries & Galloway	185	70

*death-rate is of people under 75 years old
Figure 67.5 Death-rate from heart disease in Scotland

68 Diseases of the developed world—heart disease (2)

Factors in distribution

The countries worst affected by heart disease are all in the developed world. But, as Figure 67.3 shows, there are big differences between the developed countries. Scotland has one of the worst heart disease rates in the world but, even here, some areas are much worse than others, e.g. Glasgow. The main reasons why heart disease varies so much from one area to another are as follows:

Lifestyle

In developed countries and in cities throughout the world, **the pace of life is faster**. In addition, a lot more people work in offices and take little exercise.

Diet

Some countries have healthier diets than others. For instance, the traditional Asian diet contains very little meat and dairy produce and Japanese people have a much lower heart disease rate than British people. A Mediterranean diet contains few saturated fats and people there have lower rates of heart disease.

Affluence

1. **People in richer countries can afford to eat too much** and can afford to buy cigarettes and alcohol and so are more likely to have heart disease.
2. Within developed countries, **the cheapest foods are often fatty foods** and so poorer people are more likely to develop heart disease.

Medical care

In the developed world, countries such as Australia have run very good **campaigns to educate people on how to reduce heart disease**. As a result, the disease has dropped dramatically in recent years.

Within any country, **the number of people with heart disease depends on the treatment available** locally, for example, how well the local health authorities try to diagnose and prevent heart disease and the equipment and drugs available there to treat the disease.

Controlling heart disease

Death-rate from heart disease in the UK, amongst 16–64-year-olds, has dropped by 42% in the last 10 years. Some countries, such as Australia, Canada and Sweden, have done even better than this. This has taken place because of better prevention and better treatment.

eat more	eat less
skimmed milk	full milk and cream
polyunsaturated margarine	butter
grilled food	fried food
low calorie soft drinks	milk shakes
chicken, turkey	sausages, pork pies
oats, pasta, cereals	cakes, biscuits, sweets
fruit and vegetables	chips, crisps
brown bread	white bread

Figure 68.1 Preventing heart disease

Methods of prevention

1. Better Diet

A lot of advice is now available on which are healthy foods to eat and which ones should be reduced or avoided (see Figure 68.1).

This advice has been successful in that, in the last 20 years, the amount of butter sold has dropped by 76% and the amount of full milk by 74%, while the amount of fresh fruit sold has risen by 43%. However, 45% of men and 30% of women in the UK are still overweight.

2. More exercise

Facilities for exercise have increased, e.g. jogging tracks, cycle lanes, gyms, sports centres, but there is no evidence that the average person now takes more exercise. In fact, 31% of men and 20% of women do not have enough exercise to give them protection against heart disease.

3. Reduce smoking

There have been extensive campaigns to persuade people to stop smoking, and there is more help available, e.g. nicotine patches, helplines, hypnotism. The number of smokers is now less than 20 or 50 years ago.

4. Reduce stress

People are now more aware that stress is harmful and know ways to reduce stress, e.g. relaxing by taking exercise or listening to music. But there is no evidence that stress levels are decreasing. In fact, it is more likely that they are increasing.

5. More medical check ups

As it has become more widely known that heart disease is the main cause of death, more people now have regular cholesterol and blood pressure check-ups. This should allow them to find out if they are at risk and then to take some action before it is too late.

Treatment

More equipment is being invented and used, such as pacemakers, artificial heart valves, defibrillators. By-pass surgery is steadily improving. And more drugs are being developed, e.g. aspirin to reduce blood clotting, beta-blockers to reduce heart rate, alpha-blockers to reduce blood pressure.

QUESTIONS

1 Name the five factors that affect the distribution of heart disease **(5)**

2 How does affluence affect the distribution of heart disease? **(2)**

3 Suggest why Glasgow has a higher rate of heart disease than the Scottish Highlands **(5)**

4 What evidence is there that we now eat a healthier diet? **(3)**

5 Apart from diet, describe in detail one method of preventing heart disease and describe its effectiveness **(4)**

6 Why is the treatment of heart disease so expensive? **(3)**

7* **Draw a poster to persuade people to reduce their stress levels** **(6)**

69 Diseases of the developed world—cancer (1)

Cancer is the second biggest cause of death in Britain and the whole of the developed world. It will strike 40% of the people living in this country today.

Cancer occurs when cells in the body start to divide. Unless they are stopped, the area affected will grow steadily. This can happen in many parts of the body e.g. lung, brain, stomach. So there are many types of cancer, depending upon where the disease began (e.g. lung cancer) and consequently many different causes. Figure 69.1 explains these different causes.

QUESTIONS

1 Explain how smoking causes cancer **(2)**

2 Which age-group is most likely to suffer from cancer? **(1)**

3 Name two air pollutants that can cause cancer **(2)**

4 Name two other causes of cancer **(2)**

5 *'The highest rates of cancers are in the developed world and the lowest rates in the developing world'*
Referring to Figure 69.2, how true is the statement above **(5)**

6 Describe what Figure 69.3 shows **(2)**

7 Describe the distribution of countries with the lowest rate of cancer in Western Europe (Figure 69.4) **(2)**

Look at Figure 69.5.
List the five countries in Western Europe with the worst rates of cancer in a) males, and b) females **(3)**

Causes

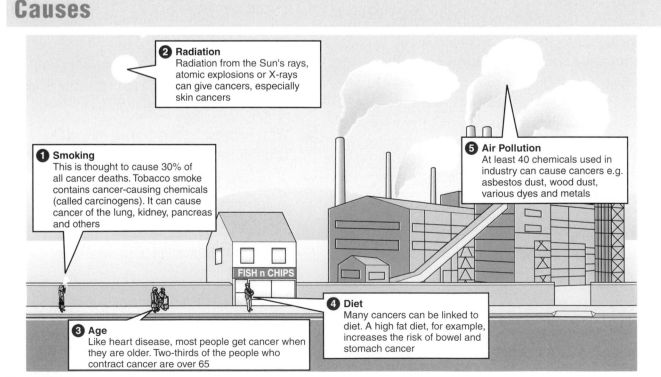

Figure 69.1 The causes of cancer

Distribution

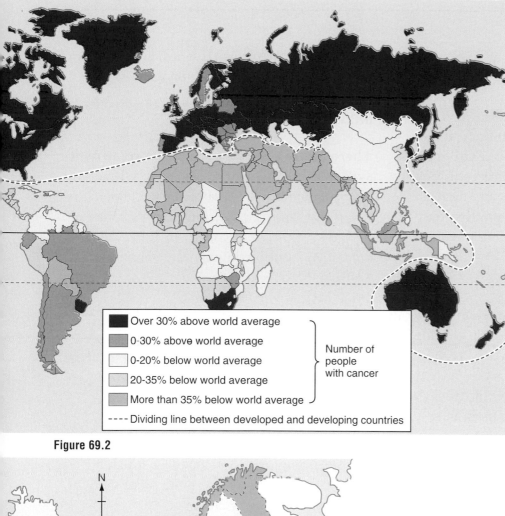

Figure 69.2

Legend:
- Over 30% above world average
- 0-30% above world average
- 0-20% below world average
- 20-35% below world average
- More than 35% below world average

} Number of people with cancer

- - - - Dividing line between developed and developing countries

	Number of cases of cancer per 100 000 males	% of people with cancer who die
Developed Countries	300	60
Developing countries	150	75

Figure 69.3

Country	Number of cases of cancer per 100 000	
	Male	Female
Austria	162	102
Belgium	195	103
Denmark	179	143
Finland	142	87
France	187	85
Germany	172	104
Iceland	131	115
Ireland	172	120
Italy	172	91
Luxemburg	195	96
Netherlands	181	107
Norway	145	103
Portugal	155	85
Spain	174	79
Sweden	123	96
Switzerland	160	93
England & Wales	164	116
Northern Ireland	166	112
Scotland	193	133

Figure 69.5

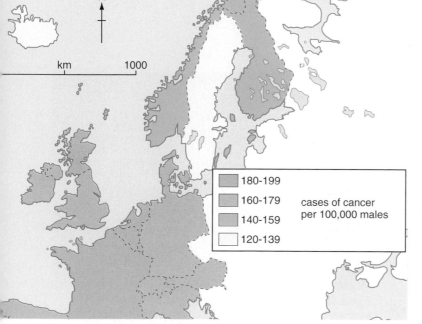

N

km   1000

Legend:
- 180-199
- 160-179
- 140-159
- 120-139

cases of cancer per 100,000 males

Figure 69.4

70 Diseases of the developed world—cancer (2)

Factors in distribution

Figure 69.2 on page 143 shows that cancer is more common in developed countries. It also shows that there are wide differences in cancer rates within the developed countries of Europe (Figure 69.4) and between males and females (Figure 69.5). The main factors affecting the distribution of this disease are the following.

Affluence

In the richer, developed countries people can afford to smoke. Many people **can also afford to go on holiday regularly to hot, sunny areas**. This makes them more likely to suffer from lung and skin cancers. In Africa, where smoking is much less common, lung cancer is very rare.

Stress

The high-pressure lifestyle more common in developed countries, and especially in cities, **leads to an increase in smoking**. People also need to go on holiday to relax and may get sunburned.

Age

Cancer is mostly a disease of older people. In the developed world, there are far more older people and so there is a greater proportion of cancer sufferers than in the developing world.

Occupation

People who **work in environments that are smoky and polluted** are more likely to contract cancer. These industrial areas are more common in the developed than the developing world.

Medical care

There are big differences in the treatment available in different areas of the UK and in different countries of Europe. For example, tablets that destroy cancerous brain cells cost £6000 for a 6-month course. In 1999, these drugs were available in only three hospitals in Scotland.

Controlling Cancer

Prevention

Screening

Screening is used to detect breast cancer before any other symptoms develop. This makes it possible to treat the disease more successfully. It reduces death-rates by 30%.

Figure 70.2 People who live longer have had more time to be exposed to cancer-causing agents and more time for slow-developing cancers to become serious, e.g. lung cancer

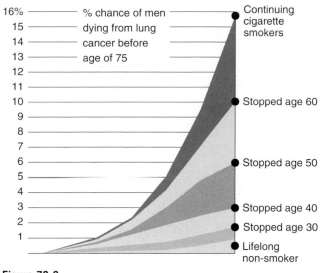

Figure 70.2

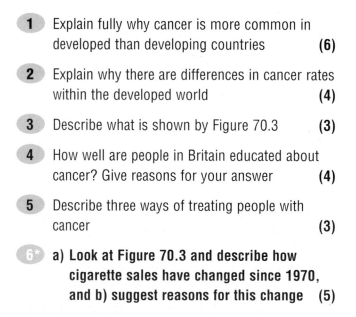

Figure 70.3

Education

If people knew more of the causes and means of preventing cancer, the disease could be reduced dramatically. Our knowledge of the disease is greater than ever before, as **health groups have run very successful campaigns to educate the public**. In 1950, for example, 80% of all men in Britain smoked and we had the highest death-rate from lung cancer in the world. In the year 2000, less than 30% of men smoked. Figure 70.2 shows how much people can reduce their chances of dying from lung cancer when they give up smoking. Similarly, people are now much more aware of the dangers of sunbathing and know how to reduce the risk of getting skin cancer. Nevertheless, experts still believe that one in three cancer deaths in Britain could be prevented.

Treatment

Unlike heart disease where cure is unusual, cancer patients are increasingly being cured. The percentage of people dying from cancer is decreasing and some experts reckon that in 50 years time the disease will be controlled.

To cure cancer, the destructive cells must be removed. This is done in three main ways:

1. **Surgery**—when the lumps or tumours are removed
2. **Radiotherapy**—when high-energy rays kill the cancer cells,
3. **Chemotherapy**—when drugs are taken that destroy the cancer cells. The first drugs were only discovered in the 1940s but they have increased and improved since that time.

QUESTIONS

1　Explain fully why cancer is more common in developed than developing countries　**(6)**

2　Explain why there are differences in cancer rates within the developed world　**(4)**

3　Describe what is shown by Figure 70.3　**(3)**

4　How well are people in Britain educated about cancer? Give reasons for your answer　**(4)**

5　Describe three ways of treating people with cancer　**(3)**

6*　a) **Look at Figure 70.3 and describe how cigarette sales have changed since 1970, and b) suggest reasons for this change**　**(5)**

71 Volcanoes (1)

Natural hazards

Natural hazards or environmental hazards are sudden events in nature that cause people problems. The problems may be slight (e.g. snow blocking roads) or severe (e.g. forest fires destroying property) or catastrophic, e.g. volcanic eruptions, earthquakes, drought, floods and tropical storms, which may kill hundreds of people.

The worst natural hazards are called natural disasters and are thought to kill 130,000 people every year, 97% of them in developing countries. It is estimated that they cause damage totalling £60 billion pounds a year. It is these, the most serious natural hazards, that are covered in this topic.

Volcanoes as natural hazards

Volcanoes find many ways of causing people problems. Volcanic ash can cover houses and streets, lava can pour out over farmland and people may be forced to leave their homes when a volcano erupts. At their worst, volcanoes are killers.

Figure 71.1 names some of the worst ones in human history.

Location of volcanoes

Figure 71.2 shows the location and distribution of active volcanoes in the world. **Active volcanoes are those that are likely to erupt**, e.g. Mt. Etna.

Notorious volcanic eruptions
Mt. Pelee (1902): 28,000 people killed by a ball of lava, that hurtled down the side of the volcano at 300 km/hr.
Vesuvius (AD 79): 2000 people suffocated by a massive downfall of hot volcanic ash that buried the town of Pompeii to a depth of 3 metres in a very short time.
Krakatoa (1883): 36,000 people killed by tidal waves up to 35 metres high.
Nevada del Ruiz (1985): 20,000 people buried by a 40 metre high mudflow (ash mixed with snow melt) sweeping down the volcano at 50 km/h, which then turned solid and trapped them.

Figure 71.1

Extinct volcanoes are those that will never erupt again, e.g. Edinburgh's volcano. There are also dormant volcanoes that have not erupted for at least 100 years, but may erupt again.

Active volcanoes are concentrated in just a few areas of the world. **Most are found near crustal plate boundaries**. In particular, they are located around the edge of the Pacific Ocean (e.g. St Helens, Fujiyama), in the middle of the Atlantic Ocean (e.g. Surtsey) and through the Mediterranean Sea (e.g. Vesuvius, Etna)

Cause of volcanoes

Mapping the distribution of a feature can often help in understanding its cause. It is no coincidence, for example, that volcanoes are found near plate boundaries.

The crust of the Earth is split into separate blocks called crustal plates. They move slowly, floating on the semi-liquid mantle

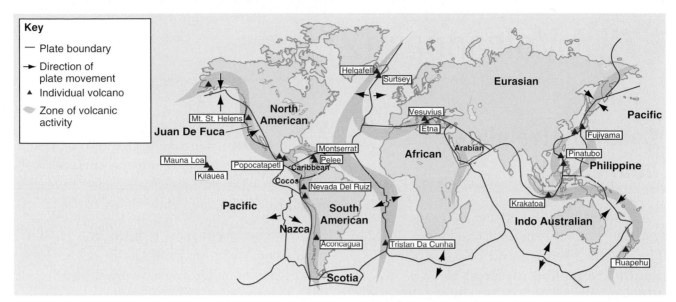

Figure 71.2 Distribution of plate boundaries and volcanoes

Figure 71.3 Vesuvius victim, found in Pompeii

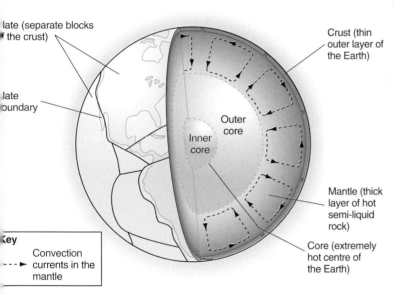

Figure 71.4 The layers of the Earth

underneath (see Figure 71.4). **Where plates meet is called a plate boundary**. There are two main types of plate boundary (constructive and destructive) and volcanoes occur at both, as is explained on the next page.

QUESTIONS

1 What is a 'natural hazard'? **(2)**

2 What is the difference between an active and dormant volcano? **(2)**

3 Describe the location of active volcanoes in the world **(3)**

4 What is a crustal plate? **(1)**

5 What is the connection between volcanoes and crustal plates? **(1)**

6 How do crustal plates move? **(2)**

Look at Figure 71.2.
   a) Name eight active volcanoes in the world
   b) For each one, name the crustal plates it is on and the one that is very near **(5)**

72 Volcanoes (2)

Constructive plate boundaries

Constructive plate boundaries are found where two plates are moving apart (see Figure 72.1). **The moving semi-liquid rocks in the mantle are pulling the crust in two different directions, so that it cracks and splits**. This **allows liquid rock from the mantle (called magma)** to rise into the crust through the cracks and **reach the surface**. When it reaches the surface it is called a volcanic eruption.

At the surface the magma is called lava. Here it cools down, turns solid and fills the crack. Then, as the plates continue to move apart, more cracks form and the process repeats itself.

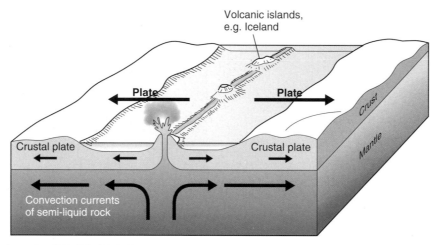

Figure 72.1 Volcanoes at constructive plate boundary

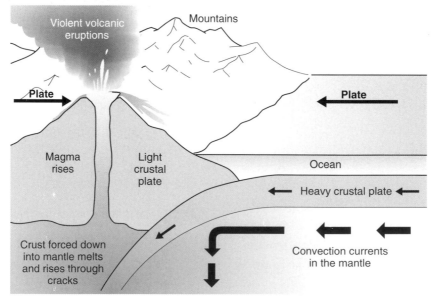

Figure 72.2 A destructive plate boundary

Features of a volcano

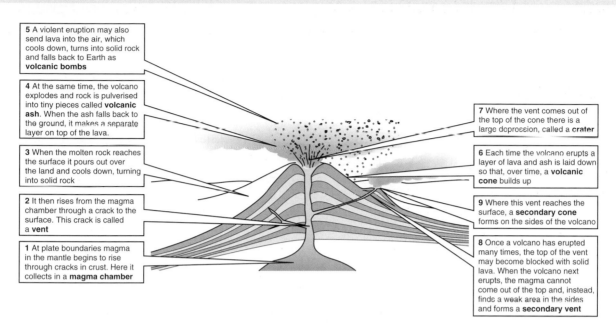

5 A violent eruption may also send lava into the air, which cools down, turns into solid rock and falls back to Earth as **volcanic bombs**

4 At the same time, the volcano explodes and rock is pulverised into tiny pieces called **volcanic ash**. When the ash falls back to the ground, it makes a separate layer on top of the lava.

3 When the molten rock reaches the surface it pours out over the land and cools down, turning into solid rock

2 It then rises from the magma chamber through a crack to the surface. This crack is called a **vent**

1 At plate boundaries magma in the mantle begins to rise through cracks in crust. Here it collects in a **magma chamber**

7 Where the vent comes out of the top of the cone there is a large depression, called a **crater**

6 Each time the volcano erupts a layer of lava and ash is laid down so that, over time, a **volcanic cone** builds up

9 Where this vent reaches the surface, a **secondary cone** forms on the sides of the volcano

8 Once a volcano has erupted many times, the top of the vent may become blocked with solid lava. When the volcano next erupts, the magma cannot come out of the top and, instead, finds a weak area in the sides and forms a **secondary vent**

Figure 72.3 The processes involved in the formation and eruption of a volcano

Destructive plate boundaries

Destructive plate margins are found where two plates move together (see Figure 72.2). **The surface rock crumple together and crack** and are squeezed up into mountains. At the same time, the heavier plate is forced down into the mantle. Here it melts and the **liquid rock makes its way through the cracks to the surface**, as a volcanic eruption. These eruptions, mixed with gases, are usually very explosive.

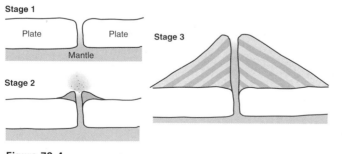

Figure 72.4

QUESTIONS

1 What is the difference between a constructive and destructive plate boundary? **(2)**

2 Explain how molten rock is able to reach the surface at a constructive plate boundary **(2)**

3 Draw a diagram to show a destructive plate boundary and label it to explain why volcanoes occur there **(6)**

4 Name the six features being described in the list below:
 a) wide vertical crack inside volcano
 b) pool of liquid rock deep in crust
 c) lava blown into the air and turning solid
 d) cone at side of volcano
 e) depression at top of volcano
 f) made of layers of lava and ash **(3)**

5* **Draw larger copies of the three stages of a volcano (Figure 72.4). Label each diagram to show what is taking place** **(6)**

The eruption of Mt. St. Helens, 1980 (1)

Cause of the eruption

Mt. St. Helens is in the Rocky Mountains near the west coast of the USA, in the state of Washington. **It lies near to a destructive plate boundary** (see Figure 73.1), where the small Juan de Fuca Plate is moving south-east and the North American Plate is moving north-west.

The small plate is being forced under the larger plate and into the mantle. Here it melts, partly because of the heat and partly because of the immense friction as two plates grind together. **As it melts, molten rock rises into the crust**. Here it builds up in magma chambers until it is able to force its way through cracks in the crust to the surface (Figure 73.2). This has happened many times, e.g. at Mt. Lassen in 1914, Mt. Rainier in 1834 and, catastrophically, at Mt. St. Helens in 1980.

The eruption

It was the 18th of May in 1980 when Mt. St. Helens erupted for the first time in 123 years. It erupted with a power 500 times greater than any atomic bomb exploded during World War Two and was the most powerful eruption on earth for the last 60 years. No lava poured out, but the eruption still had three devastating effects:

1. 400 million tonnes of **ash rose 20 km into the air**. Some rose so high, it never came down.
2. There was **a tremendous blast** from the eruption, which could be heard 300 km away.
3. **A mudflow of rock, melted ice and ash** hurtled down the mountain side at 250 km/h.

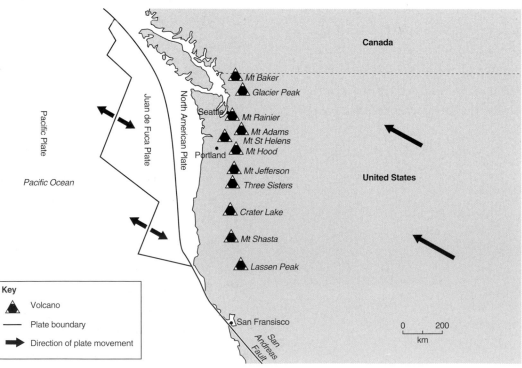

Figure 73.1 The location of Mt. St. Helens

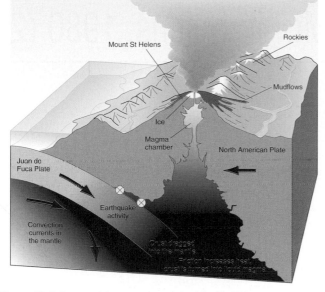

Figure 73.2 Cause of the eruption of the Mt. St. Helens

Figure 73.4 Ash clouds from Mt. St. Helens

Figure 73.5 Trees flattened by the lava flow

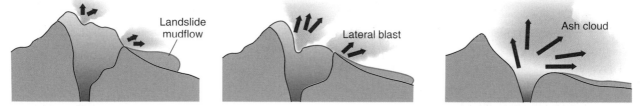

Figure 73.3 The eruption of Mt. St. Helens

Impact on the landscape

1. **The eruption of ash blew away the top of the mountain**. In seconds it changed from a mountain 2950 metres high to one that was only 2560 metres high. At the top, a crater 500 metres deep, formed.

2. **The blast killed every form of plant and animal life for a distance of 25 km** north of the volcano. Even fully-grown fir trees were flattened, up to 30 km away. About 7000 animals died, including elk and bears.

3. **The mudflow choked rivers with sediment, killing all fish and water life** and completely filling in Spirit Lake. About 12 million salmon died. The mud emptied itself into the sea at Portland, clogging up the harbour.

QUESTIONS

1. Describe the location of Mt. St. Helens **(2)**

2. When did the eruption of Mt. St. Helens take place? **(1)**

3. Give three pieces of evidence to show that the eruption was violent **(3)**

4. Describe the effects on the landscape of **a)** the ash eruption, **b)** the blast, and **c)** the mudflow **(6)**

5. With the aid of a diagram, explain why Mt. St. Helens erupted **(6)**

Imagine you are a reporter at the volcano. Write a front-page story, describing the eruption and its effects (5)

74 The eruption of Mt. St. Helens, 1980 (2)

Impact on the people

May 18, 1980 was a Sunday, so no-one was working in the forests that cover the slopes of

Mt. St. Helens. Local people had been evacuated from their homes and tourists were prevented from getting close. In spite of all this, the eruption still killed 61 people and 198 had to be rescued. Damage ran into billions of pounds.

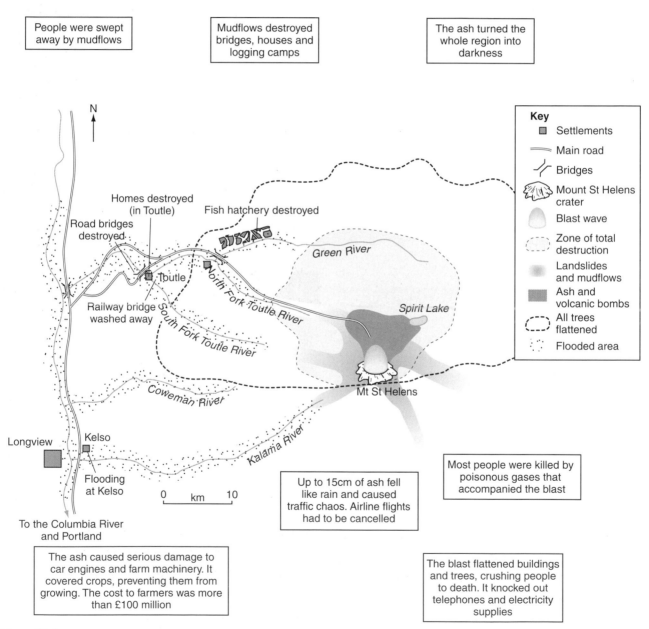

People were swept away by mudflows

Mudflows destroyed bridges, houses and logging camps

The ash turned the whole region into darkness

Key
- Settlements
- Main road
- Bridges
- Mount St Helens crater
- Blast wave
- Zone of total destruction
- Landslides and mudflows
- Ash and volcanic bombs
- All trees flattened
- Flooded area

Homes destroyed (in Toutle)

Fish hatchery destroyed

Road bridges destroyed

Toutle

Green River

North Fork Toutle River

Spirit Lake

Railway bridge washed away

South Fork Toutle River

Coweman River

Mt St Helens

Kalama River

Longview

Kelso

Flooding at Kelso

0   km   10

Up to 15cm of ash fell like rain and caused traffic chaos. Airline flights had to be cancelled

Most people were killed by poisonous gases that accompanied the blast

To the Columbia River and Portland

The ash caused serious damage to car engines and farm machinery. It covered crops, preventing them from growing. The cost to farmers was more than £100 million

The blast flattened buildings and trees, crushing people to death. It knocked out telephones and electricity supplies

Figure 74.1

Figure 74.2 Car wreck in the wake of the eruption

Figure 74.3 The blast ripped the side of the volcano off

Figure 74.4 The mudflow due to the eruption killed animals and plants

Effectiveness of the preparations

Mt. St. Helens had given clear warnings that it might erupt explosively. From March onwards there had been minor earthquakes and small eruptions of ash and steam. These gradually became more severe.

As a result, **the authorities were able to evacuate residents, tourists and forestry workers** from the surrounding area. **Emergency services were on hand**, including helicopters and aeroplanes.

But, several events upset their plans.
1. **A few people refused to leave** the area, believing they were safe.
2. Although most residents left, **scientists, reporters and cameramen all moved in**. Most of the people who died, in fact were only there because the volcano was about to erupt.
3. **The volcano did not erupt as expected**. The main damage was done by the blast which came out of the side of the volcano. This mean it affected areas beyond the official danger zone—areas where tourists were staying.

QUESTIONS

1 Describe the effects on people of:
   a) the mudflow (2)
   b) the ash eruption (6)
   c) the blast (3)

2 How did the authorities know that Mt. St. Helens was about to erupt violently? (2)

3 Describe how effectively the authorities prepared for the eruption (4)

4* **Imagine you were camping on the lower slopes of Mt. St. Helens when it erupted. Describe what you would have seen, felt and heard** (5)

75 Location and features of earthquakes

Earthquakes as natural hazards

20,000 people are killed each year by earthquakes, which makes them bigger killers than volcanoes. This is partly because earthquakes give no warning. It is also because many areas that suffer earthquakes are popular areas in which to live, e.g. California, and, in some cases, people do not know they are living in such a dangerous place. Earthquakes are also very common.

An earthquake happens somewhere in the world every two minutes. But most are very slight and they mainly occur under the sea. No-one hears of them. Sometimes, however, there are severe ones and, just occasionally, they take place under a large town. This is when earthquakes make headline news.

31.5.1970
Earthquake in Peru Causes Landslide
50,000 feared buried alive

27.3.1964
Earthquake Causes Tidal Waves 9 Metres High
port of Valdez in Alaska wrecked

18.4.1906
Fires Destroy San Francisco
caused by severe earthquake

Location of earthquakes

The location of earthquakes (Figure 75.1) is very similar to that of volcanoes (page 147). They are concentrated in just a few parts of the world. **Nearly all take place near crustal plate boundaries**.

They are particulary common, around the edge of the Pacific Ocean (e.g. Japan, California) and through the Mediterranean Sea (e.g. Turkey, Italy). Most occur under the sea (e.g. mid-Atlantic) because most plate boundaries are found there.

Features of an earthquake

An earthquake occurs when rocks inside the crust move suddenly. Where this happens is called the *focus* of the earthquake. **This sudden movement causes shock waves** to travel out in all directions. The place on the surface directly above the focus receives the worst shock waves. This is called the *epicentre* (see Figure 75.2). **There are three types of shock waves**.

1. **P waves** (push or primary waves) make the rocks move up and down—they travel the fastest
2. **S waves** (shake or secondary waves) make the rocks move from side to side—they travel at two-thirds the speed of P waves
3. **L waves** (long waves) spread out in waves along the surface—they are the slowest but the most destructive.

The **shock waves are detected on seismographs** (see Figure 75.3). The magnitude of the earthquake is **measured on the Richter Scale**. This is a logarithmic scale from 1–12. Earthquakes of scale 3 or under are minor and are not usually strong enough to be felt. Scale 6 or more are severe. No earthquake has yet registered greater than a scale 9.

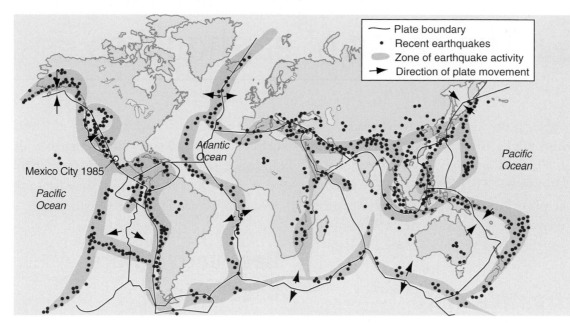

Figure 75.1 Distribution of earthquakes

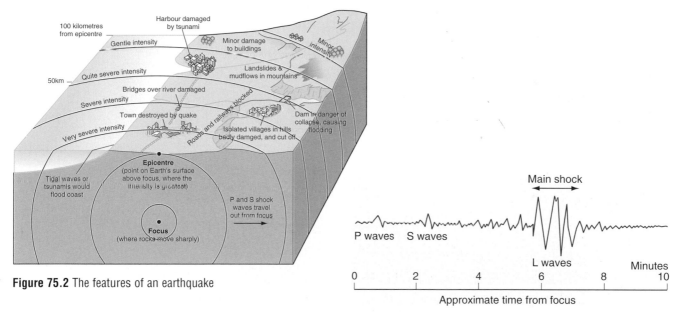

Figure 75.2 The features of an earthquake

Figure 75.3

QUESTIONS

1 Why do earthquakes kill more people than do volcanoes? **(2)**

2 Describe the location of earthquakes **(2)**

3 Explain the terms: **a)** focus, **b)** epicentre, **c)** seismograph, **d)** Richter Scale **(4)**

4 **a)** Name the three types of shock waves.

b) Which are the fastest shock waves?

c) Which are the most severe waves?

Look at Figure 75.2. List the different effects of an earthquake a) up to 50 km from the epicentre, and b) over 50 km away **(5)**

76 The cause of earthquakes

As mentioned on the last page, **earthquakes occur when rocks in the crust move suddenly**. This sets up shock waves that travel out in all directions. **This is most likely to happen at plate boundaries** where plates are trying to move in different directions. Earthquakes take place at three types of plate boundary.

puts the rocks under a lot of tension. Eventually, **some of the rocks crack and move sharply**. This causes shock waves, which travel through the crust to the surface. Here they cause the ground to shake. This is the earthquake.

Constructive plate boundaries

At constructive plate boundaries, Figure 76.1 the crust is being forced in opposite directions. This

Destructive plate boundaries

At destructive plate boundaries, Figure 76.2, **one crustal plate is being forced down below another**. But the friction between these huge chunks of crust is immense and stops the plates

Figure 76.1 Earthquake activity at a constructive plate boundary (black dot represents the focus of the earthquake)

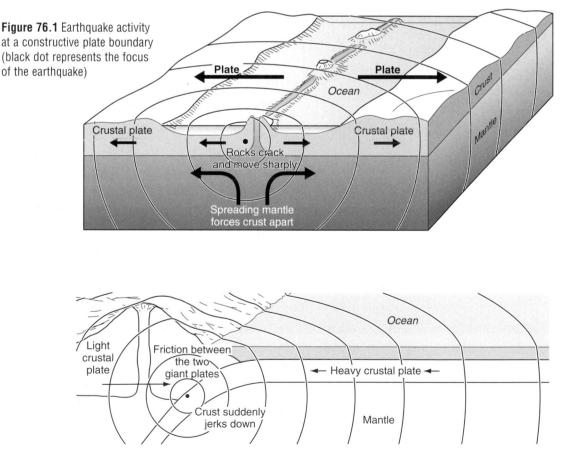

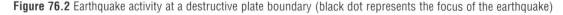

Figure 76.2 Earthquake activity at a destructive plate boundary (black dot represents the focus of the earthquake)

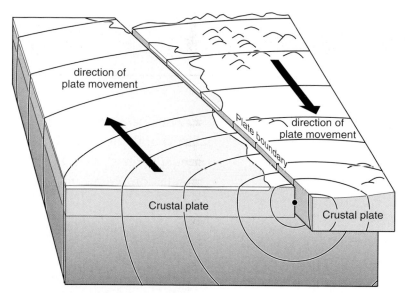

Figure 76.3 Earthquake activity at a sliding plate boundary (black dot represents the focus of the earthquake)

Figure 76.4 San Andreas fault

from moving. Eventually, however, the pressure continues to build up and **the crust jerks downwards into the mantle**. This sudden movement sends out shock waves that are felt as earthquakes on the surface, e.g. Alaska.

Sliding plate boundaries

In some areas of the world **where crustal plates meet, they just slide past one another**. No crust is destroyed or constructed and no volcanic activity takes place. But **the sliding movement is not smooth**. Because of the immense friction between the two slabs of crust, **the plates are locked together most of the time**. When the pressure has built up over a long period of time,

it is great enough to overcome the friction and this is when **one plate suddenly jerks past the other**. This causes shock waves and an earthquake on the surface, e.g. California.

QUESTIONS

1 An earthquake is caused by shock waves at the Earth's surface. What causes the shock waves? **(1)**

2 Why do areas near plate boundaries receive most earthquakes? **(1)**

3 Explain in detail why earthquakes occur at constructive plate boundaries **(3)**

4 Why do plates not move smoothly under or past one another? **(1)**

5 At destructive plate boundaries, the crust moves in a series of jerks. Explain why it moves in this way **(3)**

6 Draw an annotated diagram to explain how earthquakes occur at sliding plate boundaries **(3)**

7* **Figures 76.1, 76.2 and 76.3 show three areas where earthquakes occur.**
What are the differences between (a) Figures 76.1 and 76.2 (b) Figures 76.2 and 76.3 **(5)**

Mexico earthquake, 1985 (1)

Mexico is a developing country in Central America, south of the USA. It is located in a region where several crustal plates meet (see Figure 77.1), so earthquakes and volcanic eruptions occur regularly. Active volcanoes include Paricutin and Popocatapetl. They occasionally erupt quite violently, but rarely kill many people. Earthquakes occur more often. Mexico receives five times as many earthquakes as does the USA. Most only have minor effects, but the earthquake in Mexico in 1985 made headlines all over the world. It was the most

devastating earthquake this century but, more importantly, it took place near to the world's second largest city—Mexico City, with a population of 18 million people.

Cause

The earthquake in Mexico happened because the Cocos plate is being forced under the North American plate. The Cocos plate moves at 6 cm per year but, where it meets the other plate, **friction prevents it from moving**. As **pressure continues to build up**, however, the **friction is overcome** and, at 7:19 am on 19 September 1985, **the plate suddenly jerked 20 km down into the mantle**.

The shock waves raced outwards and reached the surface travelling at 25,000 km/h. The area where the shock waves first reach the surface is called the epicentre and this was located 50 km off the west coast of Mexico (see Figure 77.2). The shock waves registered 8.1 on the Richter Scale and **the ocean bed lurched two metres eastwards**.

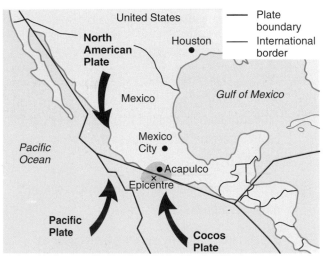

Figure 77.1 Crustal plates in Mexico

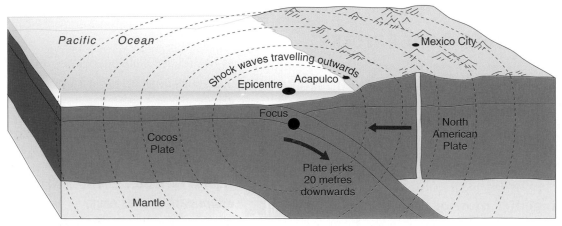

Figure 77.2 The cause of the Mexico earthquake

The earthquake lasted three minutes. That was all the time needed to reduce large areas of Mexico City to something resembling a war zone. To make matters worse, 36 hours later there were severe after-shocks, which themselves registered 7.6 on the Richter Scale.

Impact on the landscape

The earthquake devastated nearly a million square kilometres of Mexico from the west to the east coast. On the west coast, **20-metre high tidal waves smashed into hotels**, causing tourists to be evacuated. In the countryside, **villages were cut off** and unable to receive help. Their telephone lines had been cut, the roads were blocked by landslides and railway lines buckled. Some towns and cities managed to escape damage. Those that were built on rock foundations survived as the rocks absorbed most of the shock waves. But Mexico City is built on silt and mud.

Mexico City lies 400 km from the epicentre. It should not have been badly damaged. But the shock waves brought water towards the surface. This made the mud very soft so that it started to wobble and the buildings on top of it started to topple over. The vibrations had the same frequency as the natural frequency that make 9–15 storey apartments shake. These, in particular, collapsed while the tallest buildings (e.g. the Pemex Tower with 46 storeys) and even 300-year-old palaces survived. Many of the newer apartment blocks were not earthquake-proof, as building regulations had not been enforced. **Over 1000 buildings collapsed**, including hospitals, schools, apartment blocks and factories.

At the same time electricity cables snapped and gas pipes burst, which resulted in countless **fires breaking out**. These raged through the city adding to the already polluted air. **The earthquake had also severed the water pipes** so it was impossible for firemen to find enough water to put the flames out. Much of the poorer,

Figure 77.3 Landslides caused by the Mexico earthquake

inner-city areas were laid waste and there was general chaos and mayhem.

QUESTIONS

1. Which two plates caused the Mexico earthquake? **(1)**

2. Why did one of the plates suddenly jerk downwards? **(2)**

3. What evidence is there that the shock waves were very severe? **(2)**

4. Why were buildings on the coast destroyed? **(1)**

5. Why were villages cut off? **(2)**

6. Why were there so many fires in Mexico City? **(3)**

7. Apart from being burned, explain why so many buildings collapsed in Mexico City **(4)**

○ **Which areas do you think suffered most in the earthquake—mountain villages or city centres?
Give reasons for your answer** **(5)**

78 Mexico earthquake, 1985 (2)

Impact on the people

The earthquake killed approximately 20,000 people, most but not all in Mexico City. At the coast, huge tidal waves sank 30 fishing boats, 19 trawlers and four freighters and **their crews were drowned**. In the countryside, **landslides buried people alive**.

In Mexico City, **people were crushed by falling buildings**. The earthquake occurred at the start of the rush-hour, so most casualties were in apartments and few in schools and offices. Half of the city's hospitals collapsed. When the

Figure 78.1 The devastation of the Mexico earthquake

12-storey high Central Hospital collapsed, 1000 patients and staff died, yet 58 new-born babies were pulled out alive. About 60,000 people were seriously injured and some died from these injuries, partly because **they could not receive proper medical treatment** in time. **One million people lost their jobs** because the factories and shops where they worked were destroyed. The total damage came to £2.5 billion.

Despite the terrible effects, the earthquake could have been far worse. Sewage was issuing into the streets but drinking water did not become seriously contaminated and there were **few cases of cholera, typhoid and other water-borne diseases**. One-half of all Mexico's industry is found in and around Mexico City and most of it survived. The vital oilfields were not affected, and ports and airports continued to operate normally. This helped the relief effort considerably.

The relief effort: the role of the aid agencies

Short-term aid

News of the tragedy soon reached the outside world and international aid poured in.

The **Red Cross flew in doctors and nurses**. They **distributed emergency supplies—** medicines, water, clothing and tents. They took blood from volunteers and gave transfusions to those needing it.

Meanwhile, **individual countries helped with the rescue operation**. The United States sent

Figure 78.2 Emergency services helping deal with the earthquakes

Figure 78.3 Earthquake rescue worker surveys a collapsed building

Individual **countries and charities gave money** to pay for new hospitals and schools. In spite of all this help, it was to take Mexico many years to recover economically from the effects of three seismic minutes.

bulldozers, cranes and lifting gear. France sent **rescue specialists** and search dogs. Britain sent **firemen** equipped with infrared sensors to detect survivors. Mexico used 50,000 **troops**, together with police and firemen to help rescue the many thousands who were trapped. But, the whole **operation was hampered by the severe aftershocks**, which made it extremely dangerous to go into partly-collapsed buildings.

Long-term aid

After the rescue operation had been completed, it was clear that **much reconstruction had to be done**. Thousands of new homes were required. New factories, offices and shops had to be built. Broken water and sewerage pipes had to be replaced and new power lines and telephone lines had to be put in place. Yet, at the time of the earthquake, **Mexico had debts of £50 billion** and so had little money to spend on reconstruction. **The World Bank provided a loan of £200 million** to pay for rebuilding.

QUESTIONS

1 Explain why so many people died from the earthquake in Mexico City **(3)**

2 What caused deaths in other parts of Mexico? **(2)**

3 What were the main short-term tasks of the aid agencies? **(3)**

4 Who gave short-term aid and what did they give? **(4)**

5 Explain why long-term aid was needed **(5)**

6* **In a city such as Mexico City, when would be the worst and 'best' times of day for an earthquake to happen? Give reasons for your answer** **(5)**

79 Tropical storms: location and features

Location

Tropical storms are severe depressions in which **windspeeds reach over 60 km/h** but can often reach over 200 km/h. As Figure 79.1 shows, they are **found over oceans within 30 degrees of the Equator**. They start on the eastern side of oceans and move westwards, before dying out over land. **When tropical storms reach 120 km/h, they are called hurricanes**. There are local names for hurricanes in different parts of the world, as shown in Figure 79.1.

Main features

About 500 million people in 50 countries live in fear of tropical storms. They kill more people each year than earthquakes or volcanoes, yet some parts of a tropical storm are much more deadly than others. The main features of a tropical storm are listed below and shown in Figure 79.2.

1. As the storm approaches, the air pressure and temperature drop, while cloud cover and rainfall increase.
2. **Near the centre**, huge cumulo-nimbus clouds rise up, **torrential rain falls and windspeeds reach their maximum**.
3. **At the centre, the** *eye* **is calm, clear, warm and dry.**
4. After the centre of the storm, the same weather as in **2** is experienced again, with towering clouds, very heavy rain and very strong winds.
5. At the edge of the storm, the air pressure and temperature rise, while cloud cover and rainfall decrease.

Tropical storms travel at about 10 km/h, but they can speed up or slow down quickly. The route they take is called a 'track' and **they can also change direction suddenly**. On reaching coastal areas, **they can raise the level of the surface water** by up to ten metres. At high tides **this produces a storm surge**, which leads to severe flooding. Once they reach land they slow down, change direction and quickly die out. An average tropical storm lasts for 1–2 weeks.

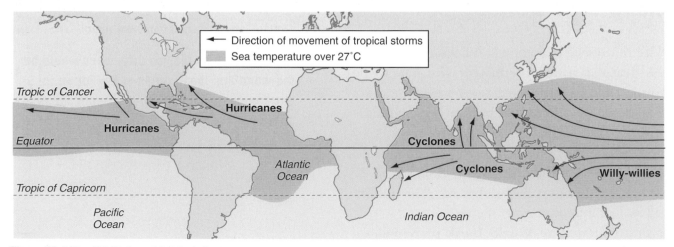

Figure 79.1 The distribution of tropical storms

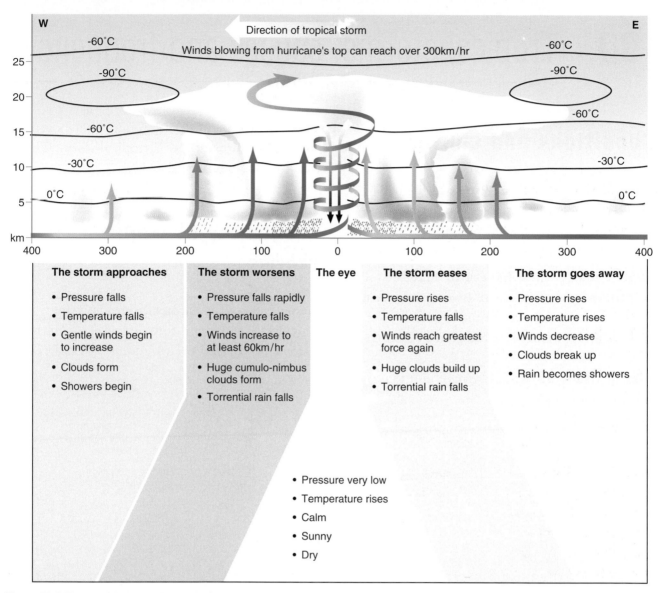

Figure 79.2 The main features of a tropical storm

The storm approaches
- Pressure falls
- Temperature falls
- Gentle winds begin to increase
- Clouds form
- Showers begin

The storm worsens
- Pressure falls rapidly
- Temperature falls
- Winds increase to at least 60km/hr
- Huge cumulo-nimbus clouds form
- Torrential rain falls

The eye
- Pressure very low
- Temperature rises
- Calm
- Sunny
- Dry

The storm eases
- Pressure rises
- Temperature falls
- Winds reach greatest force again
- Huge clouds build up
- Torrential rain falls

The storm goes away
- Pressure rises
- Temperature rises
- Winds decrease
- Clouds break up
- Rain becomes showers

QUESTIONS

1. Describe the distribution pattern of tropical storms (2)

2. What name is given to a tropical storm in **a)** the Atlantic Ocean, **b)** the Indian Ocean, **c)** China and Japan, **d)** Australia? (2)

3. **a)** Which part of a tropical storm brings the worst weather?
   b) Describe the weather it brings (6)

4. **a)** Where is the calmest weather found?
   b) Describe the weather fully experienced here (4)

5. What is a storm surge? (2)

6. Describe the movement of tropical storms (3)

The weather near the centre of a tropical storm is very different from the weather near the centre.
Describe the main differences (5)

80   Tropical storms: causes and frequency

Conditions and causes

Tropical storms are only found in certain areas of the world (see Figure 79.1, page 162). These are the areas that have the necessary conditions for them to form. They need:

1. **warm seas, which have a surface temperature of 27°C or more**, and warm water to a depth of at least 60 metres,
2. **a low air pressure**, with the air beginning to rise,
3. **damp moist air** with a relative humidity of 60% or more.

Where these conditions are found, there are five stages in the formation of a tropical storm. These are shown in Figure 80.1 below.

Frequency of tropical storms

The pictograms on the next page show the frequency of hurricane-force tropical storms in different oceans and their frequency in different months of the year.

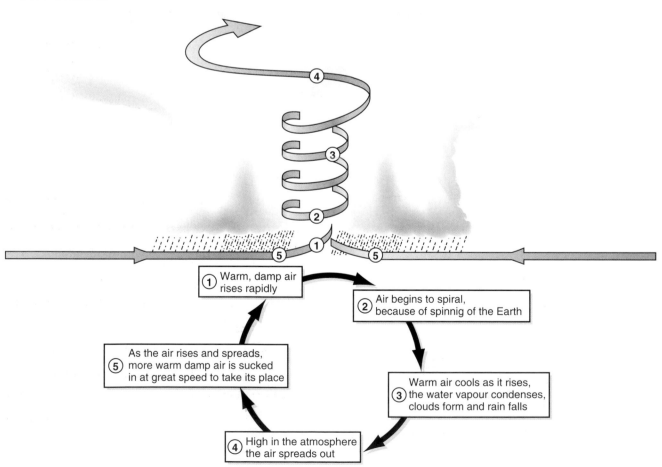

1. Warm, damp air rises rapidly
2. Air begins to spiral, because of spinnig of the Earth
3. Warm air cools as it rises, the water vapour condenses, clouds form and rain falls
4. High in the atmosphere the air spreads out
5. As the air rises and spreads, more warm damp air is sucked in at great speed to take its place

Figure 80.1 Stages in the formation of a tropical storm

Ocean area	Numbr of Hurricanes in One Year
North Atlantic	
North-east Pacific	
North-west Pacific	
South-west Pacific	
South-west Indian	
North Indian	

Figure 80.2

Frequency of hurricanes in southern hemisphere (average number in 5 years)	
November	4
December	5
January	9
February	10
March	9
April	3

Figure 80.3

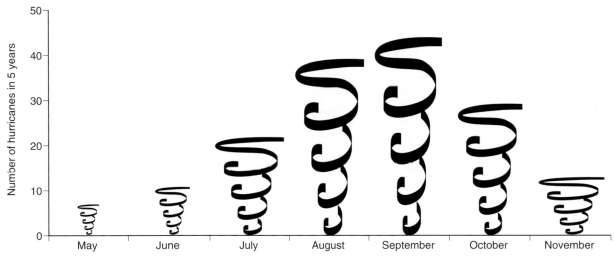

Figure 80.4 Frequency of hurricanes in the northern hemisphere

QUESTIONS

1 Describe the conditions necessary for tropical storms to form **(3)**

2 Why does rapidly rising air produce heavy rain? **(2)**

3 Why does rapidly rising air lead to very strong winds? **(2)**

4* a) Look at Figure 80.2.
Which ocean receives the most and the fewest tropical storms? **(1)**

b) Look at Figure 80.4.
Compare the frequency of northern hemisphere hurricanes in different months **(2)**

c) Look at Figure 80.3.

Draw a pictogram to show the number of tropical storms in the southern hemisphere in different months **(2)**

81 The causes of Hurricane Mitch, Central America, 1998

In October 1998, Central America made headlines all over the world when it was suddenly hit by the deadliest Atlantic hurricane in over 200 years. The cause of Hurricane Mitch can be traced back to the summertime when the hot, tropical weather caused the waters of the Caribbean Sea to become very warm. What happened next is shown in Figure 81.1 below.

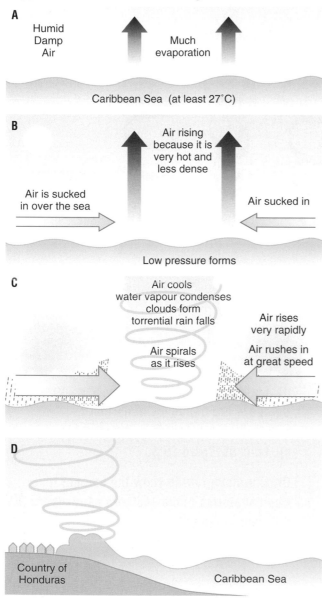

Figure 81.1

A: August/September

1. After a hot summer, **the waters of the Caribbean Sea reached 27°C.**
2. Water vapour evaporated from the sea, making **the air above humid and damp**.

B: 22 October, 1998

3. **The air** in contact with the sea **became very hot and began to rise, creating a low pressure**.
4. Air was then sucked in over the sea to replace the rising air. A tropical depression had now formed. This was the birth of Hurricane Mitch.

C: 23 October, 1998

5. By now, the hot air was rising even more rapidly. So, **air rushed in even faster, making stronger and stronger winds**.
6. As the winds reached 60 kph, Mitch became a tropical storm.
7. As the winds reached 120 kph, Mitch became a hurricane.
8. It started to move west, towards the coast of Central America.
9. On the 26th, the winds reached a peak speed of 290 kph.

D: 29 October, 1998

10. Hurricane Mitch had now reached the country of Honduras, laying waste the islands just offshore.
11. **Once over land**, without any moist air, **Mitch lost energy and began to slow down**.

The track of Hurricane Mitch

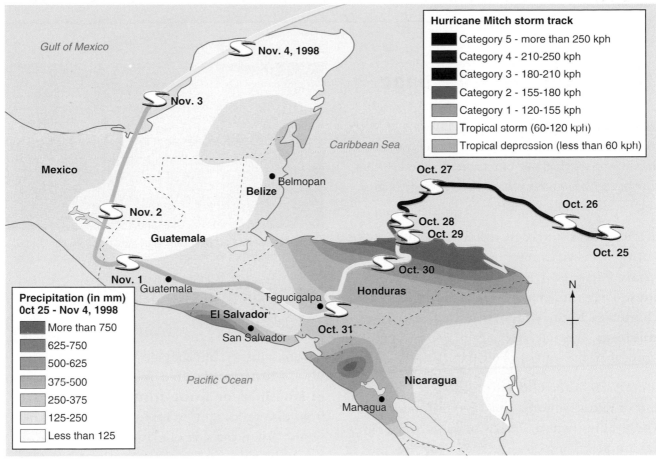

Figure 81.2

Date	location of Mitch	windspeed
October 27	Caribbean Sea	more than 250 kph
October 28		
October 29		
October 30		
October 31		
November 1		
November 2		
November 3		
November 4		

Table 1

QUESTIONS

1 Which area of the world was affected by Hurricane Mitch? **(1)**

2 Explain fully why air began to rise over the Caribbean Sea in October 1998 **(2)**

3 Explain how rising air led to the very strong winds of Hurricane Mitch **(2)**

4 At what windspeed did Mitch become **a)** a tropical storm? **b)** a hurricane? **(1)**

5 Describe the changes in Mitch's windspeed from 27 October to 2 November **(2)**

6 Why did Mitch slow down over Honduras? **(2)**

Look at Figure 81.2.
Draw Table 1 and complete it to show where Mitch was located and its windspeed between 27 October and 4 November.

82 The effects of Hurricane Mitch, Central America, 1998

Impact on the landscape

The countries worst affected by Hurricane Mitch were Honduras and Nicaragua, as they were the first countries that the hurricane reached. Other countries had similar effects, but not on such a huge scale as these countries.

In Honduras: the coast received the full fury of Mitch, with winds reaching nearly 300 kph. The **winds flattened trees, crops and power lines**. They caused **a storm surge, with waves 13 metres high, which smashed into coastal buildings**. But the effect of the rain was far worse than the winds.

Some parts of the country probably received 600 mm of rain in 6 hours and 1500 mm during the whole storm period. This is twice as much rain as Edinburgh receives in a year. No-one knows exactly because most of the rain gauges were destroyed! The **intense rain washed away soil** from the land and dumped it in rivers. Rivers grew to ten times their normal width. Every river in Honduras flooded. The power of the **floodwaters swept away 100 bridges and made 1500 km of roads impassable** (see Figure 82.4).

In Nicaragua: the hurricane was slowing down but it dropped more rain as it passed over higher ground. On the side of Casita volcano, rain formed a vast lake inside the vent, which then burst out of the mountainside. This caused a **huge mudslide** to crash down the slopes of the volcano, burying four villages. Elsewhere, **rivers flooded towns as well as countryside and left behind several metres of mud**.

Figure 82.1 Severe flooding caused by very heavy rainfall caused more problems than the winds

Impact on the people

Figure 82.2 shows the effects of Hurricane Mitch in bare statistics, but they do not tell the full story. **Most people died from being buried in landslides or from drowning**. Some died from diseases caught by having to drink polluted water. When the side of Casita volcano gave way, four villages and 2000 people were suddenly buried. **Over two million people were made homeless** throughout Central America. **Floods and high winds destroyed vast areas of crops** and this was made worse by: (a) the countries depending upon crops for their exports, (b) many of the crops being trees or bushes (e.g. banana trees), which take years to regrow, and (c) the countries being very poor with little money to repair the damage. Honduras, for example, at that time had debts of $5 billion.

The role played by aid agencies

1. The first task of the aid organizations was to **rescue people** stranded by the floods or trapped by collapsed buildings. Neighbouring

country	deaths	crop damage	other effects	cost of damage
Honduras	6500	70% of all crops destroyed and 80% of banana crop lost	20% of all the people homeless	$5 billion
Nicaragua	4000	30% of coffee destroyed and sugar harvest lost	malaria and cholera break out	£3 billion
El Salvador	200	80% of maize lost	food shortages	$1 billion
Guatemala	300	coffee and banana crops badly affected	much land unfarmable	$1 billion
Mexico and others	<100		tourists cancel holidays	<£1 billion
TOTAL	11,000			$10 billion

Figure 82.2

Figure 82.3 House destroyed in the wake of Hurricane Mitch

Figure 82.4 Bridges destroyed by flooded rivers in Honduras

countries (e.g. Mexico, USA) sent helicopters, troops and bulldozers. It was impossible, however, to save many of those trapped.

2. The second stage was to **provide emergency supplies** to the survivors. The Red Cross, for example, provided water purification kits and medical supplies to reduce the threat of disease. For the homeless, food, clothing and blankets were given. This was more successful, and there were relatively few outbreaks of disease.

3. The final stage was to **start rebuilding**. Many charities were involved in building new homes, schools, roads and hospitals. Countries donated money—Sweden $200 million, Spain $105 million and USA $80 million—but there was not enough and it did not arrive very quickly. As a result, one year later, half a million people were still living in temporary shelters, very few roads and bridges had been repaired and officials feared it could take 30–40 years for the region to recover.

QUESTIONS

1 Describe the effects of the hurricane-force winds on the landscape **(2)**

2 How did the heavy rain affect the soil? **(1)**

3 Describe the effects of the floodwaters on the land **(2)**

4 Explain how most people died **(2)**

5 Explain why the loss of crops was such a disaster for the region **(4)**

6 Describe the different purposes of the aid given **(4)**

7 Was the relief work a success? Give reasons for your answer **(3)**

8* **Make up a poster for a charity persuading people to help those affected by Hurricane Mitch** **(6)**

83 Floods as natural hazards

Location

Floods can be caused by both rivers and the sea (see Figure 83.3).

River floods occur when there is a **sudden rise in river level** and the **river overflows its banks** onto the valley floor.

Sea floods occur when **strong winds raise the water level** and the **sea covers the land** at the coast.

Causes of river floods

Figure 83.1 shows a flood hydrograph for river X. A flood hydrograph shows the changes in the amount of water in a river (its discharge) during and after a rainstorm. The peak discharge always occurs after the peak rainfall. This is because it takes time for rainwater to reach the river. **The time taken for rain to reach a river is called the basin lag time**.

Figure 83.2 shows the flood hydrograph of another river. It has a much longer basin lag time. Because of this its peak discharge is lower than that for river X and so it is much less likely to flood.

Rivers with a short basin lag time are most likely to flood. The main reasons why some rivers have a short basin lag time are as follows:

1. **the drainage basin is made of impermeable rocks**—this means the water cannot seep in but runs off quickly over the surface and into the river,

2. **there are many towns nearby**, so rain runs off over the concrete and tarmac, into drains and quickly into the river,

3. **there is little vegetation**, such as trees, to absorb the rain or slow down the run-off, and

4. **the slopes around the river are steep** so water runs off rapidly into the river.

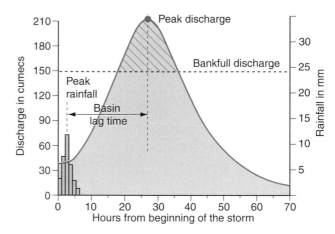

Figure 83.1 Flood hydrograph for river X

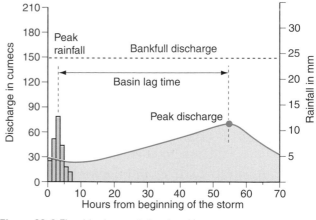

Figure 83.2 Flood hydrograph for river Y

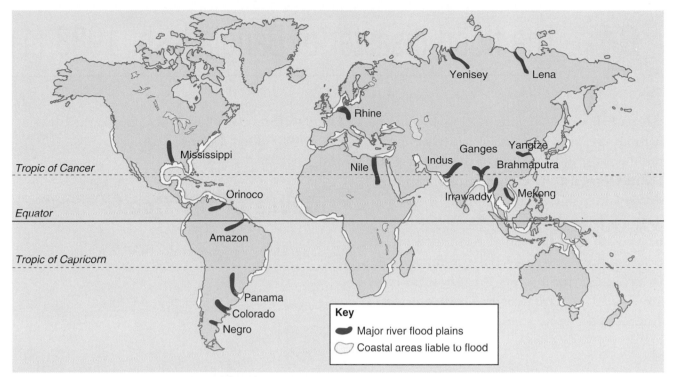

Figure 83.3 Areas in the world most liable to flood

Main features of a river flood

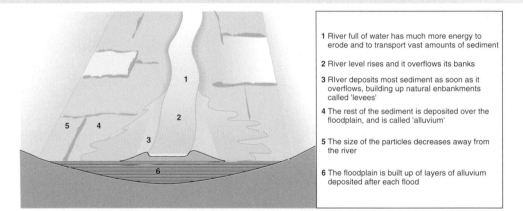

1 River full of water has much more energy to erode and to transport vast amounts of sediment

2 River level rises and it overflows its banks

3 River deposits most sediment as soon as it overflows, building up natural enbankments called 'levees'

4 The rest of the sediment is deposited over the floodplain, and is called 'alluvium'

5 The size of the particles decreases away from the river

6 The floodplain is built up of layers of alluvium deposited after each flood

Figure 83.4

QUESTIONS

1. What is the definition of 'a flood'? (2)

2. Name four large rivers which are liable to flood (2)

3. Describe the following features of a flood: flood-plain; alluvium; levee (3)

4. Look at Figure 83.4.
   Describe the changes in river discharge after a rainstorm (2)

5. a) What is 'basin lag time'? b) What is the connection between basin lag time and river flooding? (3)

6. a) In an area of impermeable rocks, explain why a river has a short basin lag time b) Describe three other reasons why rivers have a short basin lag time (10)

Describe the types of weather that cause rivers to flood (5)
(clues: waterlogged ground; a thaw)

84 The flood disaster of Bangladesh, 1988 (1)

Bangladesh is a small country in South Asia. It is half the size of the UK yet it has twice the number of people, making it the most crowded country on earth. It also suffers more natural disasters than any other country in the world. Tropical storms, droughts and, especially floods, occur with great regularity. Bangladesh suffers from sea floods when cyclones in the Bay of Bengal cause storm surges. One of these in May 1991 killed 140,000 people. The country also suffers from river floods, which was the cause of the 1988 flood disaster.

Cause of the floods

Bangladesh is prone to flooding, for three main reasons:

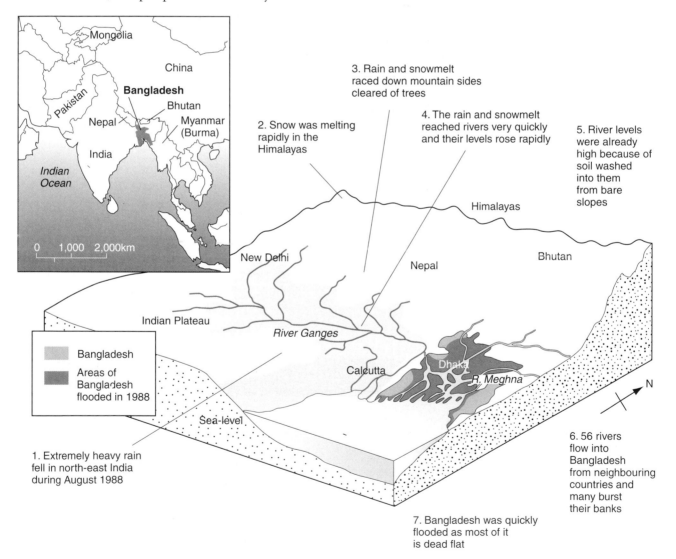

Figure 84.1 The causes of the 1988 Bangladesh floods

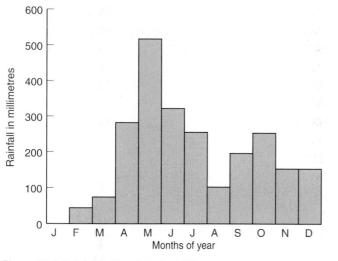

Months of year

Figure 84.2 Rainfall in Bangladesh, 1988

1. **It is very low-lying**—70% of the whole country is less than one metre above sea-level,
2. Although it is only small, **the country has 250 rivers,** the largest being the Ganges.
3. **It has a monsoon climate**, which means that for part of the year little rain falls but, during summertime, torrential downpours occur.

In 1988, three particular events led to the floods.
1. **Very heavy monsoon rains fell**, especially during August (see Figure 84.2), so the river levels rose even more quickly than usual. When the rainfall is heavy very little seeps into the ground. Instead, it runs off rapidly into the nearest river.
2. Bangladesh's largest rivers rise in the ice and snow of the Himalaya Mountains (see Figure 84.1). In 1988 **there was a rapid thaw in the mountains** and a lot of meltwater reached the rivers and raced down to Bangladesh and the sea.
3. During the years before the floods **people of the Himalayas had been cutting down the forests** at a very rapid rate. The country of Nepal had cut down half of its forests in 30 years. With fewer trees, rain reaches the rivers more quickly. Also, with fewer trees, soil erosion has increased by 400 times. Much of the soil is washed into the rivers where it sinks to the bottom. As a result, the beds of the rivers were rising by as much as 5 cm per year.

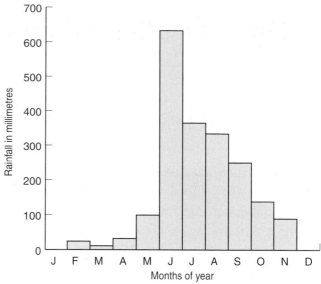

Months of year

Figure 84.3 Rainfall in north-east India, 1988

QUESTIONS

1 Describe the physical landscape of Bangladesh (mention height and rivers) **(2)**

2 Describe the climate of Bangladesh **(2)**

3 What is the connection between flooding in Bangladesh and snow melting in the Himalayas? **(2)**

4 Explain why cutting down forests in the Himalayas has increased flooding in Bangladesh **(4)**

5 Look at Figures 84.2 and 84.3. Compare the rainfall in Bangladesh and north-east India in 1988 **(3)**

6 *'The rainfall in Bangladesh in 1988 does not explain our floods during August and September.' (Bangladeshi climatologist)*
a) Explain what the statement above means **(2)**
b) What were the main reasons for the Bangladeshi floods in 1988? **(6)**

7* **List five different reasons why Bangladesh had floods in 1988** **(5)**

85 The impact of the Bangladesh floods, 1988 (2)

Impact on the landscape

By August 1988 the rivers flowing into Bangladesh were full of water from the heavy rains and the snow-melt and were also choked with soil from the mountainsides. By mid-August floods covered 20% of Bangladesh. This was normal. The people use flooded fields in which to plant rice. The floods also leave behind silt, which increases the fertility of the soil. But, in late August, sudden, intense rainstorms hit north-east India. Vast amounts of rain rushed over the surface into the rivers crossing the border into Bangladesh. The worst floods in living memory hit this small, developing country. The effects were devastating.

- **75% of the whole country disappeared under water** (see Figure 85.1)
- in the countryside only the tops of trees could be seen
- in the cities, water was over two metres deep
- **7 million houses were destroyed or damaged**
- nearly 1000 bridges collapsed
- **25% of the crops were lost**, especially rice
- the country remained underwater for six weeks

Figure 85.1 The flooded countryside in Bangladesh

impact of the floods	
area flooded (km^2)	84,059
deaths	2379
crop damage (million hectares)	7.16
cattle lost	172,000
houses destroyed or damaged (millions)	7.2
hospitals and clinics flooded	1445
schools flooded	8481
roads damaged (km)	68,827
railway lines damaged (km)	638
bridges damaged	922

Figure 85.2 Statistics from the floods

Impact on the people

The floods killed over 2000 people, mostly by drowning. It was almost impossible to save stranded people as the roads, railways and airports were all under water. **Water-borne diseases killed many people**, especially diarrhoea, cholera and dysentery, because people were forced to drink the polluted floodwater. At the height of the floods, 500 new cases of cholera were being reported daily. Anti-diarrhoea drugs were available but could not reach the people that needed them. Some people even died from snake bites, as both snakes and people sought safety in dry areas. Some might have been saved by medical treatment but **most of the hospitals and health centres were also under water.**

Those who survived had lost everything. Harvests were ruined and there were **food shortages**. Food stocks were already low as they had been used following the severe floods in 1987. About **200,000 cattle also lost their lives**, increasing the hardship to farmers.

A longer-term problem was the **enormous cost**

Bangladesh (1988)	
% of people who can read	30
TV sets per 1000 people	4
radios per 1000 people	40

Figure 85.3 Warning people about floods is difficult in Bangladesh

of rebuilding the country. Table 85.2 shows the total damage, which amounted to £600 billion. To pay for this **the government had to use money that would otherwise have been spent on improvements to farming and services**.

The floods did, at least, make the soil very fertile again. But this was small compensation. Disasters such as these floods make it impossible for Bangladesh to improve its standard of living. The floods of 1988 came just a year after other severe floods which were themselves the worst in 70 years.

Predicting the flood

During the summer of 1988, **people in Bangladesh were kept informed of floods by the Flood Forecasting and Warning Centre** (FFWC) in the capital, Dhaka. It records rainfall and river levels and also forecasts the weather using satellite images of clouds and pressure systems. It then uses computers to work out when and where the rivers are likely to flood.

When the severe storms occurred at the beginning of September, **the FFWC did issue warnings** on radio, television and in the newspapers. **They advised people to move to higher ground** and to take emergency supplies with them. **The FFWC were also able to inform the emergency services**, so they were ready to help.

But **they could not issue warnings in time**. It only took two days for the heavy rains in India to cause the flooding Bangladesh. **The floods**

Flood warning

Yesterday all major rivers in the country recorded rises in water levels and remained above danger level.

The Ganges is likely to rise by 10 to 20 cm at all places while the river Brahmaputra is likely to rise by 20 to 25 cm and cross the national record.

The river Buriganga at Dhaka and the Kumar at Faridpur may cross the danger level and the flood situation in Dhaka and Narayanganj is likely to worsen.

Issued by the Flood Forecasting And Warning Centre (27 August, 1988)

Figure 85.4

were also deeper and covered a much greater area of the country than the FFWC predicted. Even the higher ground was not safe. Also, as can be seen from Figure 85.3, **many people did not know of the flood warnings**. Even if people had found out, it is difficult to know what they could have done in the short time before the deadly floods struck.

QUESTIONS

1 How much of Bangladesh suffered flooding and to what depth? **(1)**

2 Describe the effects of the flooding **a)** on the countryside, **b)** on the cities **(4)**

3 Explain why so many people died in the floods **(5)**

4 Explain how the floods affected the development of the country **(3)**

5 Flood warnings were given on television, radio and in newspapers. Do you think this was the best way of informing people? Give reasons for your answer **(3)**

6 **a)** In what ways did the FFWC help people?
b) In what ways did the FFWC fail? **(8)**

Make up a newspaper front page describing the floods **(5)**

86 Drought

Location and distribution of droughts

A drought is a **long period of time when little rain falls**. This is **unexpected** and is **below average for that area**.

Figure 86.1 shows the areas most prone to droughts. It does not include deserts, as these areas expect to have little rain, so the dry weather is not below average. The areas with droughts have the following distribution:

- nearly all are within 40 degrees latitude of the equator
- few are very close to the equator
- all are in areas that receive an average of less than 500 mm of rain each year
- most are next to hot deserts

It is in these areas where many people live because there is enough rainfall *on average* for them to farm. But, unfortunately, there are many years when rainfall is well below average.

Main features

When water on the surface heats up it evaporates into the air as a gas (water vapour). When trees and vegetation give off water into the air as a gas, it is called transpiration.

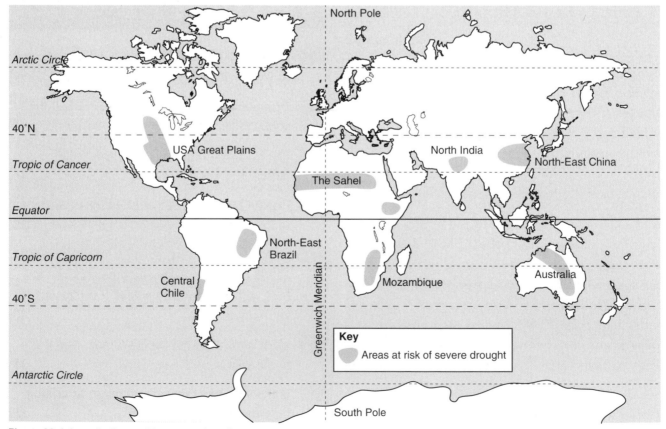

Figure 86.1 Areas in the world prone to droughts

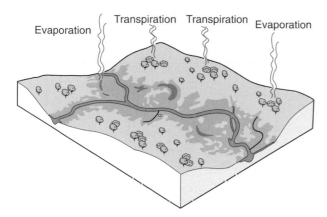

Figure 86.2

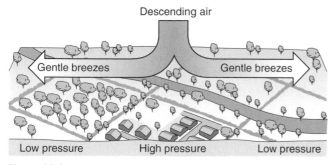

Figure 86.3

The main features of a drought are:

1. **evaporation and transpiration are greater than the rainfall,** so that
2. **the soil becomes drier,** so that
3. **less vegetation and fewer crops grow**.

The causes of drought

Droughts occur for two reasons.

1. **The area has high pressure conditions for a long period**.

In a high pressure or anticyclone, the winds are very light and the air is descending (see Figure 86.3.) For rain to fall, air has to rise and the water vapour cool down and condense. When the air is descending the weather is dry. If the high pressure remains for a long time, a drought will occur.

2. **The area is affected by offshore winds for a long period**.

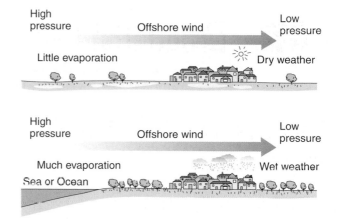

Figure 86.4

Offshore winds blow overland. This means they cannot pick up much evaporated water, so they bring dry weather. If an area receives offshore winds for a long period, it results in a drought. Winds blow from high to low pressure, so when a high pressure is centred over dry land the winds blowing from it will be offshore (see Figure 86.4). If, on the other hand, the high pressure is centred over water, the winds blowing from it will be onshore and will bring rain.

QUESTIONS

1 What is meant by a drought? **(2)**

2 **a)** Name four areas in the world that often suffer droughts **b)** Describe the distribution of drought areas in the world **(5)**

3 'A drought occurs when evaporation and transpiration are greater than the rainfall.' Explain what the statement above means. **(2)**

4 What pressure conditions bring drought? Explain why **(4)**

5 What type of winds bring drought? Explain why **(4)**

6* **Look at Figure 86.4. What are the differences between the two diagrams?** **(5)**

87 Drought in Central Sudan, 1984–1990 (1)

Sudan is a huge country in Africa, ten times the size of the UK. There are many environments here, ranging from rainforest in the south to the Sahara Desert in the north. The area that suffers from unexpected droughts is central Sudan, just south of the Sahara Desert. This is part of the Sahel zone, which stretches from coast to coast across Africa. It is also, unfortunately, the area where most people in the Sudan live.

Central Sudan usually receives about 400 mm of rain a year and nearly all of this falls between June and September (see Figure 87.2). Since 1968 the rainfall has been mostly below average but, in

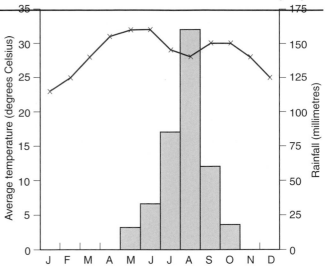

Figure 87.2 The climate of central Sudan

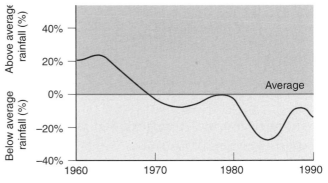

Figure 87.3 Rainfall in central Sudan: difference from the average

1984, the drought became much more severe. Over the next 6 years the region received only 75% of its normal rainfall (see Figure 87.3).

Causes of the drought

Central Sudan usually receives rain between June and September. This is because it receives onshore winds at that time but, **between 1984 and 1990, the summers were affected by offshore winds** (see Figure 87.4). These were north-easterlies, which had blown over the Arabian Desert and brought very hot, dry, dusty conditions.

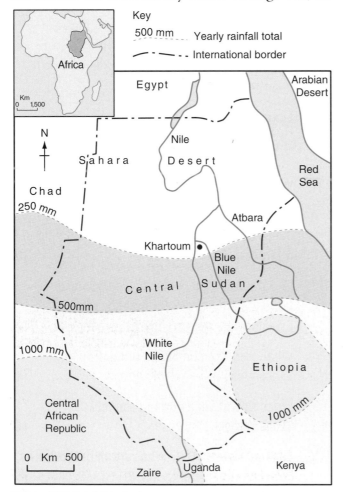

Figure 87.1 Rainfall in Sudan

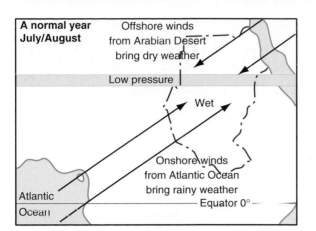
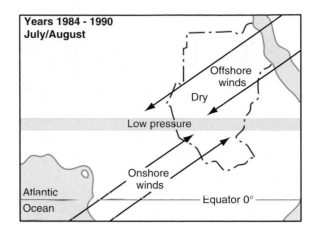

Figure 87.4 Causes of drought in central Sudan

Central Sudan usually has onshore winds because they blow from the Atlantic Ocean to the low pressure over northern Sudan. But, **between 1984 and 1990, the low pressure only reached as far as southern Sudan**. This meant that the onshore winds did not reach the central region. Instead, the winds came from the north-east and the area remained dry. With very little rain falling at other times of the year, the region began to experience a serious and long-lasting drought.

Impact on the landscape

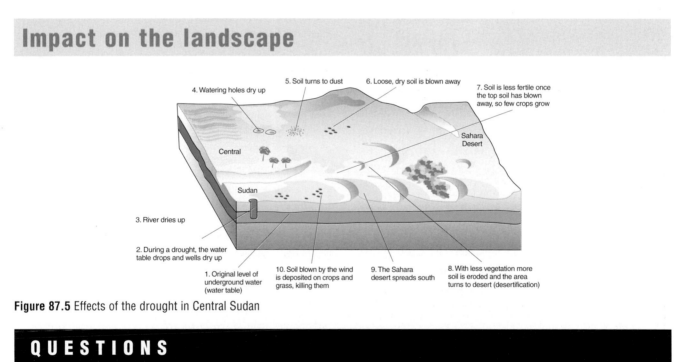

Figure 87.5 Effects of the drought in Central Sudan

QUESTIONS

1 When is the rainy season and dry season in central Sudan? **(2)**

2 Look at Figure 87.3. Compare the rainfall in central Sudan between 1960 and 1990 with the average rainfall **(3)**

3 Between 1984 and 1990, **a)** why did the winds not bring rain to central Sudan, **b)** why did the winds not come from the south? **(4)**

4 Explain why soil is easily eroded during a drought **(4)**

5 What is meant by 'desertification'? **(2)**

Look at Figure 87.5. Write down five ways in which this area will be different after several months of rain **(5)**

88 Drought in central Sudan, 1984–1990 (2)

Most people in central Sudan are nomadic herders. They own cattle, sheep, goats and camels, which they graze beside the rivers in the dry season as this is the only area where there is water and grass. In the wet season they move away from the rivers as there is grazing over the whole region at this time. Some of the people are also farmers. There is just enough rain in an average year for them to grow crops such as millet.

After the first year of drought in 1984, **the crops did not grow** and **there was not enough grazing** for their animals. The people borrowed money and sold some of their livestock to survive. After two years of drought, the people had to borrow more and sell more animals. By now, because **there was so little grain its price was much higher**, making it even more difficult for people to afford. After a few years the farmers and herders had nothing left. **A famine spread across the region as the people became hungry and weak. Many died.** Those who survived had no choice but to emigrate. **Some migrated to cities**, such as Khartoum, which could not cope with all the extra people. **Most migrated to refugee camps.** Here some food was available but, because the camps were so overcrowded, there were water shortages and outbreaks of disease. The refugees were also trapped. They no longer had seeds or animals so **they could not return to their land once the drought ended**.

The role of aid agencies

With so many people dying from the effects of hunger and disease, **the first task of the aid agencies was to deliver emergency aid**. This

Figure 88.1 The human face of the drought in Sudan

Figure 88.2 Appeal for help after the Sudan drought

was chiefly **food and medicines**. It was important to get the food to the villages quickly, before people started to move to the refugee camps. Aircraft were much more useful in doing this than were trucks.

Aid agencies also tried to provide long-term aid for the region. They believed it was better to spend some money trying to reduce the chances of another famine in the future. They started schemes that were **designed to stop the spread of the desert**. Some of these schemes are explained in Figure 88.3.

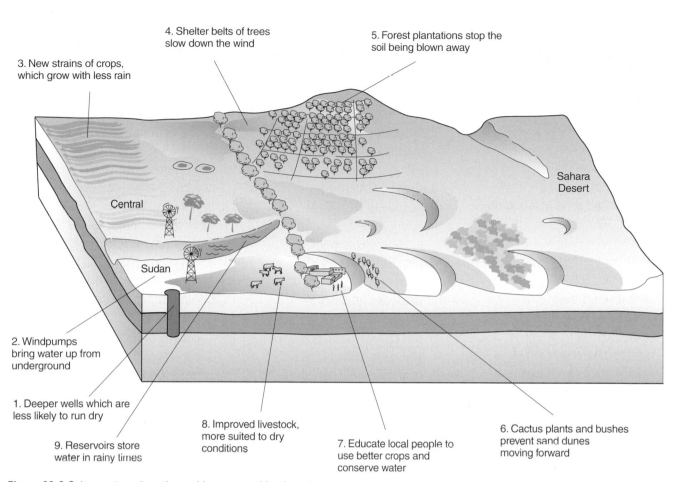

3. New strains of crops, which grow with less rain

4. Shelter belts of trees slow down the wind

5. Forest plantations stop the soil being blown away

Sahara Desert

Central

Sudan

2. Windpumps bring water up from underground

1. Deeper wells which are less likely to run dry

9. Reservoirs store water in rainy times

8. Improved livestock, more suited to dry conditions

7. Educate local people to use better crops and conserve water

6. Cactus plants and bushes prevent sand dunes moving forward

Figure 88.3 Schemes to reduce the problems caused by drought

Figure 88.4 Schemes such as digging deeper wells were initiated to try and prevent further drought in the future

QUESTIONS

1 Explain why so many people died during the drought in the Sudan **(4)**

2 To where did some people migrate? **(2)**

3 What problems did people face once they had moved? **(2)**

4 Describe the emergency aid provided by aid agencies **(2)**

5 Choose three of the long-term aid schemes shown in Figure 88.3. Explain in detail how each helps to reduce the effects of a drought **(6)**

6* **List the different ways in which the drought affected the people of the Sudan (5)**

Index